ACTA UNIVERSITATIS UPSALIENSIS
Uppsala Dissertations from the Faculty of Science and Technology
83

To Marlene,
from the author

Anna T. Danielsson

Doing Physics – Doing Gender

An Exploration of Physics Students' Identity Constitution in the Context of Laboratory Work

UPPSALA
UNIVERSITET

Dissertation presented at Uppsala University to be publicly examined in Polhemssalen, Ångströmlaboratoriet, Lägerhyddsvägen 1, Uppsala, Friday, April 24, 2009 at 10:15 for the degree of Doctor of Philosophy. The examination will be conducted in English.

Abstract

Danielsson, A T. 2009. Doing Physics - Doing Gender. An Exploration of Physics Students' Identity Constitution in the Context of Laboratory Work. Acta Universitatis Upsaliensis. *ACTA UNIVERSITATIS UPSALIENSIS Uppsala Dissertations from the Faculty of Science and Technology* 83. 270 pp. Uppsala. 978-91-554-7454-6.

In Sweden today women are greatly under-represented within university physics and the discipline of physics is also symbolically associated with men and masculinity. This motivates in-depth investigations of issues of physics, learning and gender.

This thesis explores how physics students' simultaneously constitute the practice of physics as enacted in student and research laboratories and their physicist identities in relation to this practice. In particular, it focuses on how these constitutions can be understood as gendered. Previously, physics education research has often limited 'gender perspective' to focusing on comparisons between man and woman students, whereas this study conceptualises gender as an aspect of social identity constitution. A point of departure for the thesis is the theoretical framework which combines situated learning theory and post-structural gender theory. This framework allows for a simultaneous analysis of how students 'do physics' and 'do gender', thereby making a theoretical contribution to physics education research.

In the empirical study twenty-two undergraduate and graduate physics students were interviewed about their physics studies, with a particular focus on laboratory work.

The analytical outcomes of the study illustrate a wide variety of possible identity constitutions and possible ways of constituting the physicist community of practice. For example, the students expressed conflicting interpretations of what are suitable practices in the student laboratory in terms of the value of practical versus analytical skills. The boundaries of the physicist community of practice are constituted in relation to, for example, other disciplines, interdisciplinary practices and a traditional femininity practice. Thus, the thesis demonstrates the complexity in physics students gendered negotiations of what it can mean to be a physicist.

The ambition of the thesis is further to promote discussions about gender and physics, by engaging readers in critical reflections about the practice of physics, and, thus, to inform the teaching practice of physics.

Keywords: Physics, Learning, Higher Education, Science Education, Physics Education, Gender, Situated Learning

Anna T Danielsson, Department of Physics and Materials Science, Box 530, Uppsala University, SE-75121 Uppsala, Sweden

ISSN 1104-2516
ISBN 978-91-554-7454-6

urn:nbn:se:uu:diva-98907 (http://urn.kb.se/resolve?urn=urn:nbn:se:uu:diva-98907)

Printed in Sweden by Geotryckeriet, Uppsala 2009.
Distributor: Uppsala University Library, Box 510, SE-751 20 Uppsala
www.uu.se, acta@ub.uu.se

Till min familj och mina vänner

Contents

PART II *Conceptual Framework*

Populärvetenskaplig sammanfattning

Att bli fysiker.

Genusperspektiv på fysikstudenters identitetsskapande i relation till arbete i laboratoriet.

I den här avhandlingen utforskar jag, från ett genusperspektiv, hur universitetsstudenter i fysik lär sig att bli fysiker – hur de skapar sig en fysikeridentitet. Detta är således en tvärvetenskaplig avhandling, i skärningspunkten mellan fysik, utbildningsvetenskap och genusvetenskap. Här i sammanfattningen kommer jag samtidigt som jag skissar huvuddragen i avhandlingen, peka vidare mot de avhandlingsdelar som kan intressera olika läsare.

Avhandlingstiteln ”Doing Physics – Doing Gender”, som är en engelsk ordlek på både kön/genus och fysik som göranden, kan väcka en rad frågor: Vad har fysik med kön/genus att göra? Hur kan kön/genus vara något som görs? Och, eftersom detta är en fysikdidaktisk avhandling, hur relaterar detta ”görande” av fysik och kön/genus till lärandet av fysik? Nedan behandlas dessa frågor kortfattat.

En anledning till att fysiken är intressant ur ett genusperspektiv är att genus där samtidigt är både synligt och osynligt. Å ena sidan ses ämnet i sig ofta som helt opåverkat av sociala strukturer, å andra sidan är mansdominansen inom fysiken överväldigande. Jag diskuterar i avhandlingen hur fysiken kan förstås som präglad av genus, inte bara på grund av mansdominansen, utan också på ett symboliskt plan. Till exempel delas vetenskaperna ibland in i hårda och mjuka, bland naturvetenskaperna uppfattas fysiken som hårdare än biologin, som därmed anses passa bättre för kvinnor, medan den hårdare fysiken förknippas med män (Benckert 2005). Fysikens genusladdning diskuteras närmare i bland annat avsnitt 8.6 jag och inkluderar också fysikens genusladdning i mina analyser. Detta är en av de saker som skiljer min avhandling från tidigare fysikdidaktisk forskning med genusperspektiv. Sådan forskning har ofta sett genus som något begränsat till de individuella studenterna, vanligtvis jämför dessa studier kvinnors och mäns prestationer eller deras ageranden i klassrummet.

Jag arbetar i avhandlingen med teorier både om lärande och om genus. Teoretiskt tar avhandlingen sin utgångspunkt i situerad lärandeteori och poststrukturalistisk genusteori (dessa teorier introduceras kort nedan och förklaras närmare i kapitel 3).

Att lära sig fysik kan innebära en rad olika saker. Den första tanken som kommer upp är kanske att lära sig ämnesstoffet, till exempel om studenter förstår skillnaden mellan värme och temperatur. Det finns också gott om fysikdidaktisk forskning som fokuserar just på denna typ av begreppsmässig förståelse (se kapitel 2). Men att lära sig fysik handlar inte bara om att lära sig ämnesstoff, det handlar om att tänka, agera och tala som en fysiker – i korthet, att lära sig delta i en fysikerkultur. Inom situerad lärandeteori brukar man kalla en sådan "kultur" en praktikgemenskap. Genom att delta i praktiken både formar vi våra egna identiteter och praktikgemenskapen som sådan, vi både påverkar den och påverkas av den, och det är detta som inom situerad lärandeteori kännetecknar lärande (mer om detta i avsnitt 3.1). Från ett situerat lärandeperspektiv är alltså nyckeln till att vara och att bli en fysiker deltagandet i en praktik – det räcker inte att hävda att man är en fysiker, man måste också kontinuerligt "bevisa" detta genom att delta i, och också förhandla, fysikerpraktiken. Att tänka sig att man, för att uppfattas som "fysiker", behöver iscensätta "fysiker" på ett korrekt sätt är antagligen ganska okontroversiellt. Att tänka på kön på detta sätt kan dock vara mer provocerande, men detta är precis vad poststrukturalistisk genusteori utmanar oss att göra. Mycket förenklat kan man säga att kön inom denna teoretiska strömning ses som något som "görs", inte något man bara "är" eller "har", även om detta görande sker i förhållande till de ramar som sätts bland annat av kropp och samhälle. Som Elvin-Nowak och Thomsson (2003) skriver: "Att analysera kön utifrån ett verbperspektiv är att rikta strålkastaren mot människors aktiva handlingar – de handlingar som vi alla är inbegripna i hela tiden, i alla relationer och i alla situationer. För kön finns överallt, inom oss, runt oss och mellan oss." (s. 11). Mer om detta i avsnitt 3.2.

Min avhandlings teoretiska bidrag är att jag kombinerar situerad lärandeteori och poststrukturalistisk genusteori i mitt teoretiska ramverk, för att jag ska kunna analysera studenters samtidiga "görande av fysik" och "görande av kön" (se kapitel 4).

Det datamaterial som avhandlingen bygger på är intervjuer med studenter som läser till fysiker vid universitetet. Jag har intervjuat grundutbildningsstudenter om deras erfarenheter av arbetet i kurslaboratoriet samt examensarbetare och doktorander om deras arbete i forskningslaboratorier. I analysen av intervjuerna använde jag mig av de teorier jag skissade ovan. I kapitel 6 beskriver jag datainsamlingen och i appendix C och D finns mina intervjuprotokoll. Mer om teori och metod finns i del två av avhandlingen, men om det är resultaten av själva intervjustudien du är mest intresserad av rekommenderar jag dig att gå direkt till de analytiska resultaten i del 3 (kapitel 7-9).

I "Praktiken i fysiklaboratoriet: Individuella perspektiv" (kapitel 7) fördjupar jag mig i 13 av intervjuerna. Här fokuserar jag på hur studenterna talar om arbetet i fysiklaboratoriet. Hur ska man agera i laboratoriet? Vad är viktigt att vara bra på? Inte viktigt att vara på bra? Vad anser de själva att de är bra på? När det gäller examensarbetarna och doktoranderna intresserar jag mig också för hur de beskriver "fysikern" samt hur de beskriver övergången från student till forskare.

I "Skapandet av en fysikerpraktikgemenskap: Kollektiva perspektiv" (kapitel 8) utgår jag från hur de intervjuade studenterna som grupp beskriver fysikerpraktiken och vad det innebär för dem att vara fysiker. Bland annat beskriver studenterna två olika sätt att vara fysikstudent på i laboratoriet, en mer teoretiskt intresserad student, som fokuserar på analyserande och diskussioner, kontra en mer praktiskt intresserad student, som fokuserar på experimentutförandet. I grundutbildningsstudenternas beskrivningar ställs dessa två sätt att vara fysiker på delvis i konflikt med varandra, medan doktoranderna och examensarbetarna snarare beskriver en mångkunnig fysiker som behärskar både analyserandet och experimentutförandet. I kapitlet diskuterar jag också de olika gränsdragningar studenterna gör i förhållande till "fysikern"; sker gränsdragningen i förhållande till andra discipliner eller i förhållande till äldre, mer erfarna kollegor? När blir man tillräckligt erfaren för att få kalla sig fysiker? De olika sätten att vara fysiker på analyseras också i förhållande till studenternas görande av kön. Till exempel kan den praktiska och den analytiska fysikerstudenten sägas vara representanter för olika typer av maskulinitetspraktiker, en slags arbetarklassmaskulinitet fokuserad på praktiska färdigheter och en slags akademisk medelklassmaskulinitet fokuserad på ett rationellt, analytiskt tänkande. Det blir dock också tydligt att det finns normer för hur en kvinna som läser fysik förväntas vara, normer som studenterna både omförhandlar och ibland helt vänder sig emot.

I "Identitet i praktiken: Att bli fysiker?" (kapitel 9) står de individuella studenterna återigen i centrum, men här fördjupas analysen med hjälp av mitt teoretiska ramverk och med hjälp av slutsatserna från kapitel 8. I detta kapitel fokuserar jag på hur studenterna skapar sig en fysikeridentitet i relation till fysikerpraktikgemenskapens normer och gränser, och hur dessa normer och gränser också förhandlas av studenterna. Min analys visar hur de studenter som identifierar sig som fysiker också är de som vågar göra dessa förhandlingar, till exempel genom att omdefiniera gränserna på ett sådant sätt att de själva inkluderas. Gränserna kan vidare sättas på en mängd sätt, mot andra vetenskaper, mot tvärvetenskaplig praktik, mot kvinnlighet eller mot den praktiska fysikerstudenten. För flera av de intervjuade kvinnorna är gränsen mellan fysiken och vad som ses som en traditionell kvinnlighet tydlig. Genom att framställa sig som annorlunda än "traditionella kvinnor" kan de finna en plats i den mansdominerade fysiken. Kapitel 9 utgör kärnan i min analys och ska du bara läsa ett avhandlingskapitel är det detta jag rekommenderar.

Även om avhandlingsstudien kan beskrivas som grundforskning, och som sådan driven av nyfikenhet och en önskan att förstå frågeställningar kring fysik, lärande och genus, så är den också ett bidrag till diskussionen om kvinnors underrepresentation inom naturvetenskap och teknik. Som jag diskuterar i avsnitt 10.2 hoppas jag också att avhandlingen ska kunna inspirera fysiklärare och fysikstudenter till kritiska reflektioner kring genus och fysik.

PART I

Introducing and Situating the Research

CHAPTER 1

Introduction

> I fell in love, simultaneously and inextricably, with my professors, with the discipline of pure, precise, definite thought, and with what I conceived of as its ambition. I fell in love with the life of the mind. I also fell in love, I might add, with the image of myself striving and succeeding in an area where women had rarely ventured.

The words above are from Evelyn Fox Keller (1977), describing her experiences as a woman studying physics in the 1950s. In many ways her story captures essential aspects of this thesis, both in regard to the motivations for the thesis (personal as well as professional) and, as we shall see, in regard to the methodology. But that is stepping ahead in my introduction. This is a thesis about physics students and how they, in their learning to become physicists simultaneously, 'do physics' and 'do gender'; as Evelyn Fox Keller describes in the narrative above. Her story is for all means and purposes a gendered story about doing physics; about how she constitutes herself as a particular kind of woman and a particular kind of physicist. Noticeably, how the words of Evelyn Fox Keller sharply contrast the general portrayal, in education research and the public debate, of women and physics as 'incompatible' – regardless of whether this incompatibility is framed in terms of neurobiology and gender-related spatial abilities, or in terms of women's wish for a more socially relevant science. Seldom are the voices of women finding pleasure in doing physics heard. In this thesis you will meet women and men who in various ways find pleasure in doing physics, but you will also see their struggles in reconciling the doing of physics with other aspects of themselves.

Evelyn Fox Keller is also able to eloquently convey, using only a few lines, how studying physics goes far beyond learning (or not learning) of the content matter. It is clear from her description that the learning of physics cannot be understood as an isolated activity, but needs to be understood in a broader personal and societal context. In the words of Brickhouse (2001):

> Learning is not merely a matter of acquiring knowledge, it is matter of deciding what kind of person you are and want to be and engaging in those activities that make one part of the relevant communities.
>
> (Brickhouse 2001, p. 286)

Thus, the learning of physics needs to be understood as an activity that is deeply entangled in a complex conglomerate of the personal, the social, the political, the scientific and so on; involving everything from the reactions you get when you introduce yourself as a physics student to a new acquaintance, to the way you engage in dinner-time discussions about nuclear power. Therefore, a more complete understanding of students' learning of physics must expand the meaning of 'learning'; must dare to ask questions about what consequences the doing of physics has for a student's life outside the classroom; what identities are available for a particular student; how a student's participation in physics is intertwined with their participation in other social contexts; what a student communicates by studying physics. But not only that, the studying of physics is in itself a highly complex activity involving aspects such as problem-solving, group work, examinations and lectures. Furthermore, what makes physics particularly important for my study is its grounding in experimental work. As a physics student you spend a considerable amount of your time in the laboratory, where you are expected to acquire a wide variety of knowledge and skills, both in relation to concepts and methods. I find the complexity of 'learning in the laboratory' fascinating and the possibility for students to constitute different identities in relation to this complex activity made me focus my investigation on this particular part of their physics education.

Finally, I must add that I also find Evelyn Fox Keller's account highly intriguing from a personal perspective; as a woman physicist I could immediately recognise myself in her words. In fact, when first reading her description I was surprised by the extent to which Fox Keller was able to put *my* experiences into words. It is my hope that the 'student stories' in this thesis will provide similar experiences for its readers, whether it is through thought-provoking recognition or the illumination of completely new perspectives.

1.1 Purpose of the Thesis

Continuing in the spirit of the introduction, I am interested in how students learn to become physicists, with a particular focus on the gendering of this. As explained in the introduction I explore the issue in relation to the students' participation in laboratory work. In the introduction I also argued that it is important to conceptualise learning as something more than the mere acquiring of content knowledge. The perspective employed in this thesis is that learning can be understood as the constitution of an identity. In the case of physics, not only does one learn the subject matter of the discipline, one also learns to become a physicist, to participate in the disciplinary culture of physics. How one participates in physics is further related to whom one sees oneself as being and becoming in that context. From this perspective gender becomes relevant, not as a way of sorting students into categories of men and women, but as one aspect of this identity constitution (this is further elaborated in Chapter 3).

Exploring the learning of physics in terms of the constitution of gendered identities has significance from several different perspectives. Firstly, it contributes to an intra-disciplinary development of physics education research (see Chapter 2). Secondly, it contributes to an advancement of the teaching of physics (see section 10.2). Thirdly, it is relevant from a societal perspective; in a broad sense, this thesis can be seen against the backdrop of societal concerns about the decline in young people's interest in science[1] and in particular women's under-representation in physics[2].

The concern with women's under-representation has given rise to numerous programs designed to attract more women into science and technology, as well as international organisations working with the issue.[3] However, despite all efforts, women are still under-represented in physics, and a critique that has been made against programmes aiming to attract more women to science and technology is that they are working from a premise that women's choices are constrained by a lack of information about scientific and technological work (Henwood 1996).

In this thesis I turn towards the students, women and men, who have enrolled in university physics, for an investigation of how they constitute their

[1] See, for example, ROSE (The Relevance of Science Education) (http://www.ils.uio.no/english/rose/index.html)

[2] Among the total number of students at university in Sweden about 60% are women. However, among the students enrolled in undergraduate physics in 2006/07 only 29% were women and in Engineering Physics only 20% were women (Statistisk centralbyrån, 2008a). Among the physics professors in 2007 less than 7% were women. (Statistiska centralbyrån, 2008b)

[3] Two examples of programmes designed to attract more women to science and technology are WISE (Women into Science and Engineering) in the U.K. (see Henwood, 1996, 1998) and NOT-projektet in Sweden (Gisselberg, Ottander, and Hanberger 2004) An example of an international organization working with issues of gender, science and technology is GASAT (Gender And Science And Technology), which organizes regular conferences, see http://www.gasat-international.org.

chosen discipline and their own identities in relation to this discipline, its norms and expectations. The purpose of the thesis is to *explore how physics students' simultaneously constitute the practice of physics as enacted in student and research laboratories and their physicist identities in relation to this practice. In particular, the thesis focuses on how these constitutions can be understood as gendered.*[4] Empirically the research purpose was approached through interviews with university physics students and the outcomes of this empirical study are found in Chapters 7 to 9. In order to adequately explore the research purpose, theoretical development has also been required. Here the theoretical purpose of the thesis became *to formulate a framework that allows for the exploration of physics students' gendered identity constitution in, and beyond, my empirical study.* This framework is presented in Chapter 4.

1.2 The Study Context

The research was carried out at an old, traditional university in Sweden, henceforth referred to as 'the University'. The University is a well-established research university, and can be characterised as being elite. The physics research prides itself with its long-lasting traditions and has traditionally been primarily centred around experimental physics. The University's physics research is considered a high status activity, both within and outside the University. Today physics research is carried out at several different departments, in several sub-disciplines of physics, covering a wide array of theoretical as well as experimental physics. There is both basic research and applied research, sometimes carried out in collaboration with industry.

There are several undergraduate programmes that include physics. The physics stream of the Master of Science programme and two of the engineering programmes are strongly focused on physics, but there are several other engineering programmes where physics makes up a smaller part.[5]

The first year of the physics stream of the Master of Science programme is largely devoted to mathematics courses, but over the course of the programme the proportion of physics courses increases. As the programme pro-

---

[4] This research purpose involves many interrelated components and as such cannot unpack into specific research questions without loosing the complexity I am aiming to capture in my study.

[5] All the students I interviewed were enrolled in pre-Bologna physics education, and this section consequently describes the University's physics education as it was organized prior to the Bologna process. The Master of Science programme (Naturvetarprogrammet) is a four year degree programme, resulting in a Swedish 'magister examen'. In this thesis I have chosen to label the students enrolled in the Master of Science programme prior to their Master's research project 'undergraduate students'. The students doing Master's or PhD projects are called 'graduate students'.

gresses the students' freedom in choosing courses also increases; one can specialise in, for example, meteorology, astronomy, solid state physics, molecular physics, or theoretical physics. However, there is also the possibility of putting together one's own combination of courses. During the first year of study approximately one third of the students are women, and among the students who in their third year specialise in 'physics' (i.e. not meteorology or geophysics) approximately one quarter are women. Among the physics professors at the University less than ten percent are women. About half of the beginner students at the Master of Science programme are over twenty years of age, which is a slightly higher proportion than in the engineering programmes. A year cohort of students specialising in physics consists of roughly ten to fifteen students. The attrition rate of the Master of Science programme is high; around fifty percent of the students drop out.

The Master of Science programme is a study intense program, with several hours of scheduled teaching each day. Outside the scheduled hours the students are expected to write laboratory reports, read the literature and work with problem-solving. Much of this takes place in study groups, formed by the students themselves. Apart from the study-oriented activities, there are also plenty of social activities organised by and for the physics students. Consequently, the physics students spend a considerable amount of their time together.

The teaching of the undergraduate students is carried out in rather traditional forms: Lectures in large lecture halls are combined with problem-solving sessions. Many of the physics courses also include a laboratory course; approximately five laboratory exercises, which the students work on in pairs. In contrast to lectures and problem-solving sessions the laboratory exercises are compulsory and last for about four hours each. The students take at least two courses in parallel, typically four courses per semester. The courses are assessed through written exams (in the middle and at the end of the semesters) and by completion of the laboratory course.

The last semester of the programme is devoted to a Master's research project. The project is an individual research project that can be carried out either in a research group at the university or in industry. During the Master's research project the student takes part in the daily work in a laboratory (or similar) and under the supervision of researchers, the student plans and carries out a project linked to the group's research. The Master's research project is finalised through the completion of a Master's thesis.

1.3 Structure of the Thesis

The thesis is divided into four major parts.

In Part I, consisting of Chapters 1 and 2, my study is introduced and situated within previous research. So far I have given the introduction, giving the background to my research and situating and presenting my research purpose. **Chapter 2** situates my study within previous research.

Part II presents the theoretical and methodological framing of the study and consists of Chapters 3-6. In Chapters 3-5 I present the conceptual framework of my research. **Chapter 3** introduces the broader theoretical staging of the research, that of situated learning theory and post-structural gender theory, which in **Chapter 4** is focused into my theoretical framework. In **Chapter 5** I introduce the analytical tools that I used as a bridge between the empirical material and the theoretical framework. **Chapter 6** is concerned with methodological considerations; here I discuss the data collection, issues of trustworthiness and ethics, as well as outline how the analysis was carried out.

In Part III the outcomes of the analysis of the empirical material are presented in Chapters 7, 8 and 9. In **Chapter 7** I begin the presentation of my analytical outcomes, through individual student narratives. In **Chapter 8** the presentation of the analytical outcomes continues; here the focus is on the students interviewed as a collective group as a means to explore the physicist community. The presentation of the analytical outcomes is completed with **Chapter 9**, where the full conceptual framework is employed for a deepened analysis of the students from Chapter 7.

Finally, Part IV consists of the concluding discussion, with **Chapter 10** bringing the thesis to a close; the outcomes of the study are discussed and based on them recommendations for further research and for teaching are discussed.

CHAPTER 2

Literature Review

2.1 Introduction

In the following chapter educational research relevant for my research is summarized and discussed. As such, the aim of the chapter is to situate my study within previous research.

The chapter begins with an overview of physics education research. This section of the literature review is made up of two quite dissimilar parts. The first part of the physics education section is a general, and relatively chronological, overview of the field. It could be characterised as being rather descriptive; the focus of this overview is on research trends and theoretical developments, aiming to illustrate how physics education research over the years has progressed and broadened. In the second part I focus on research about gender and physics education. These articles are critically examined and categorised with regard to their view of gender and of physics. Following the overview of physics education research, educational research with a gender perspective in science and technology is presented. Here a brief historical overview is given and then particular attention is paid to research dealing with identity constitution.

2.2 Overview of Physics Education Research

2.2.1 Introduction

In a broad sense much of the existing physics education research (PER) deals with students' understanding of physics and is aimed at informing teaching and curriculum design for improving learning outcomes (Redish 2003; Thacker 2003). This interest stems from both a concern that traditional teaching methods might not be the most effective for teaching physics to an increasingly diverse student body, as well as a concern about the decline in students choosing to study physics at university level (Thacker 2003; van Aalst 2000). Early work in PER grew out of university physics; concerned by the fact that many physics students seemed to emerge from physics teaching with substantial gaps in their understanding of physics, physicists began to conduct studies of the teaching and learning of physics. These studies were, due to the researchers' background in physics, largely a-theoretical (McDermott and Redish 1999). Later, with inspiration from studies in general science education as well as fields such as ethnography and psychology more theoretical developments within PER started to emerge (see, for example, diSessa 1993 and Redish 1999). Methods used have typically been questionnaires and/or interviews (van Aalst 2000).

McDermott (1991) write that PER's most significant impact on instruction came from the need for a greater focus on the student in both teaching practice and curriculum design. In particular, transmission-based epistemology and its associated practice have been shown to be relatively ineffective for optimizing learning. Building on forms of constructivism it was argued that students need to construct their own knowledge and in this construction it is important that the knowledge the students already have is taken into account. While research with this constructivist perspective is still used a great deal in PER, Heron and Meltzer (2005) point out that aspects of PER have also advanced well beyond documenting the shortcomings in student learning and of traditional methods of instruction – as the following literature review will show.[6]

2.2.2 Students' Conceptions

One of the major trends in PER has been the investigation of students' so called naïve understandings of the physical world and how those understanding differ from those of the physics discipline. These student understandings have been characterised as, for example, misconceptions, alternative conceptions and alternative frameworks. The more systematic investigations of students' 'misconceptions' in physics began in the late 1970s; Warren (1979) summarized some difficulties student had with understanding the concept force and also suggested some pedagogical implications. Later Helm (1980) described a number of 'misconceptions' in various fields of physics among South African students. Two early seminal papers dealing with students' understandings of Newtonian mechanics are Clement (1982) and McCloskey (1983). Clement was able to show how many physics students possess stable conceptions regarding the relationship between force and ac-

[6] Also see the PER resource letters by McDermott and Redish (1999) and Thacker (2003).

celeration. His conclusion is that, 'apparently one cannot consider the student's mind to be a "blank slate" in the area of force and motion' (p. 70). McCloskey (1983) carries the argument forward stating that people, based on their everyday experiences, form well-articulated theories of motion, that can be best characterised as a 'naïve impetus theory'. However, later research questioned whether students' ideas are consistent enough to be viewed as naïve 'theories'. Halloun and Hestenes (1985a, 1985b) could, for example, conclude that students seemed to possess a mixture of concepts and that they were inconsistent with their applications of such concepts. Finegold and Gorsky (1991) also reached a similar conclusion, with the exception that they found some consistency in students' conceptions regarding forces acting on objects in motion.

The work on students' conceptions also helped to give rise to an influential model for learning called 'conceptual change' (see, for example, Posner et al. 1982). The basic idea in conceptual change is that a person exchanges an existing conception for a more suitable alternative conception by coming to understand how this alternative conception is more intelligible, plausible and/or fruitful than the existing conception (Hewson 1982). Duit and Treagust (2003) describe how this is usually done in practice:

> The classical conceptual change approach involved the teacher making students' alternative frameworks explicit prior to designing a teaching approach consisting of ideas that do not fit the students' existing ideas and thereby promoting dissatisfaction. A new framework is then introduced based on formal science that will explain the anomaly.
>
> (Duit and Treagust 2003, p. 673)

The conceptual change model has been extensively debated, developed and criticized. For example, from a physics perspective, it was challenged by Linder (1993) who argued that it is inadequate to portray meaningful learning as a change of conceptions. Since, without consideration of the context even many physics conceptions cannot be viewed as 'correct' or 'incorrect', thus the notion of conceptual change as a model for learning needs to be understood in terms of changing one's relationship with the context.

Initially, most of the research on students' conceptions was situated in mechanics, but since then there has also been an expansion into other areas, such as thermodynamics (for example, Yeo and Zadnik 2001), optics (for example, Ambrose, Shaffer, Steinberg, and McDermott 1999; Colin and Viennot 2001; Goldberg and McDermott 1987; Singh and Butler 1990), mechanical waves (for example, Chu, Treagust, and Chandrasegaran 2008; Wittmann 1999), electromagnetism (for example, Maloney, O'Kuma, Hieggelke, and Van Heuvelen 2001; Tsai, Chen, Chou, and Lain 2007), special relativity (for example, Hewson 1982; Scherr 2007) and quantum mechanics (for example, Domert, Linder, and Ingerman 2005; Mashhadi 1994; Müller and Wiesner 2002; Petri and Niedderer 1998).

In summary, '[a]mong those who follow or participate in science education research, it has become standard to accept that students come to courses with conceptions that differ from scientists' and must be addressed in instruction' (Hammer 1996, p. 1319). How this ought to be done in practice is, however, a highly debated question. One approach that has been used to address misconceptions in learning is the 'elicit, confront, resolve approach', where a conceptual conflict between a widespread misconception and the corresponding expert conception is generated, which the students then, are required to resolve (Shaffer and McDermott 1992). The research on student conceptions has thus given rise to the development of teaching methods (see section 2.2.5) and also the development of theories of learning (see section 2.2.3).

2.2.3 Development of Theories of Learning

As pointed out by Smith et al. (1993-1994) much of the research into students' conceptions has been largely a-theoretical; aiming to describe students' conceptions rather than developing theoretical frameworks to relate students' conceptions to their learning. Smith et al. furthermore criticize much of this work for its lack of developing mechanisms for change of conceptions. In short, they consider the depiction of a 'misconception' as something that needs to be confronted and replaced as being inconsistent with a constructivist perspective on learning. Within the constructivist perspective of learning the focus is on how more advanced knowledge states (for example, expert understanding of physics) are contiguous with prior knowledge states (for example, novice understanding of physics). Consequently Smith et al. (1993-1994) argue that there are more similarities between expert and novice understandings of physics than first is apparent. For example, novices do use highly abstract entities in their reasoning about physics problems and naïve physical conceptions do continue to play an important role in experts' reasoning. Thus, novice and expert reasoning differ more in quantity than in quality, and what will 'shift' as a novice moves to a more expert understanding of physics is not the concepts themselves, but the contexts wherein they are applied. In other words, misconceptions are characterised as 'faulty extensions of productive prior knowledge' (Smith et al. 1993-1994, p. 152). diSessa (1993) makes a similar argument in his 'Towards an Epistemology of Physics'. At the heart of his argument is the view that novice physics learners' ideas about the physics world do not constitute an organized structure. Instead, he argues that novice physics learners possess a set of loosely connected ideas that are evoked in particular situations. He refers to these constructs as phenomenological primitives (p-prims). P-prims are, according to diSessa (1993), based on experience (thus, the name) and linked to specific phenomena. In our learning of physics these p-prims become refined, not replaced. Here Hammer and Elby (2002) point out:

> The ontology of p-prims has several advantages over the ontology of conceptions. First, it provides theoretical structure to account for the sensitivity to context of students' reasoning, as different p-prims are more or less likely to be activated in different circumstances. Second, it provides an account of productive cognitive resources from which students may construct more adequate understanding.
>
> (Hammer and Elby 2002, p. 178)

Hammer et al. (2004) have, using a cognitivist approach, developed diSessa's (1993) ideas. They argue that conceptions are too large a cognitive unit for understanding students' learning and suggest an approach based on the idea of the more fine-grained 'resources'[7]. A resource, for example, could be 'more effort implies more result' or an intuitive sense of 'balancing'. Thus, resources cannot be thought about as correct or incorrect (as in the case with 'misconception'), but a key to an expert understanding of, for example, physics is to apply the appropriate set of resources for a given context. Consequently, learning is described 'not as the acquisition or formation of a cognitive object, but rather as a 'cognitive state' the learner enters or forms at the moment, involving the activation of multiple resources' (Hammer et al. 2004, p. 5). Hence, a crucial aspect in Hammer et al.'s (2004) view of teaching is one of helping students to gain knowledge of the cognitive resources they already have and to be able to apply these appropriately across different contexts. This could be characterised as a metacognitive teaching approach.

In summary, there has been a move from viewing students' ideas as problematic misconceptions that need to be confronted and replaced to a constructivist based view of them as resources for learning that can be developed through teaching.

2.2.4 Contemporary Directions in PER

Early research within PER was, as I have pointed out, largely focused on students' (mis)conceptions. However, the scope of PER has broadened markedly and I will, in the following, introduce some contemporary, important strands of research, with a particular focus on the research of domains epistemology, metacognition, and representations.

2.2.4.1 Student Epistemology

One of the more important theoretical areas of growth in PER is in epistemology. I will primarily focus on two different types of research within this domain. Firstly, on research seeking to build a cognitive model for students'

[7] They use the term resources as a generic term for p-prims and epistemological primitives (see section 2.2.4.1)

epistemologies. Then secondly, on research focusing on the relationship between students' epistemologies and their approaches to learning.

Elby and Hammer, applying the resource perspective described earlier in relation to students conceptions, have developed a cognitive model for students' epistemology (Elby and Hammer 2001; Hammer and Elby 2002). Their starting point is a critique of what they claim to be a general consensus among the majority of researchers examining student epistemology, namely that students, for example, ought to understand science as 'fundamentally tentative and evolving rather than certain and unchanging' (Elby and Hammer, 2001, p. 555). Their argument is that this sort of claim is far too general to be helpful for better understanding of student learning. Thus 'epistemological stances' need to be understood as context dependent and, further, that it is important to distinguish between the correctness and the productivity of epistemological beliefs. For example, viewing knowledge as tentative rather than certain is neither productive nor correct across all contexts. Viewing Newton's laws as certain might, for example, be productive for introductory physics students, but not for more advanced physics students (Elby and Hammer 2001).

The projected context-dependentness of epistemological beliefs is further elaborated on by Hammer and Elby (2002). They argue that instead of viewing epistemologies in unitary terms, they should be viewed as consisting of epistemological resources that are neither correct nor incorrect, but which need to be applied in their appropriate contexts.

Building on previous research in PER as well as fields such as neuroscience and sociolinguistics, Redish (2004) formulated a suggestion for an overarching theoretical framework for understanding students' learning of physics, that included notions of conception and epistemology. The framework can be seen to be rooted in research on human cognition. Redish (2004) describes the core of his theoretical framework as follows:

> My theoretical framework describes student knowledge as comprised of cognitive resources in various forms and levels of hierarchy. Within each level is a collection of resources that are primed, activated, and deactivated depending on context and control.
>
> (Redish 2004, p. 16)

Thus, both in dealing with students' conceptions and their epistemologies Redish models them in terms of resources. Within a certain context, a certain frame, a number of associated resources will be activated. Hence, learning physics is largely about 'reorganizing' the students' existing resources. Consequently, for a teacher it is of crucial importance to 'frame' not only the actual problems in a way that activates the appropriate resources, but also to 'frame' the learning situation in a way that activates the most useful epistemological resources. This idea is further developed by Redish (2004) as he discusses possible implications of his framework both for instruction and for research.

A. science as discovery or 'knowing about the world'
B. science as accumulation of fact and explanations of how things are and how they work
C. science as a process of enquiry undertaken by 'scientists'
D. science as an accessible way of looking at the world and, as such, part of everyday life
E. science has a social dimension
F. science as empowering

Table 1. Views of science expressed by the students in the study by Linder and Marshall (1998)

Another strand of research on student epistemologies has focused on the relationship between students' epistemological stances and their approaches to learning (see, for example, Linder and Marshall 1998; Ryder, Leach, and Driver 1999). Linder and Marshall's (1998) starting point is an introductory physics course designed with the purpose of making students become independent, lifelong learners, through developing students' reflections on their own learning by creating a metacognitive epistemological framing. In their study, in the beginning of the course all students were voicing relatively unsophisticated views of learning as well as of science, characterised as belonging to categories A, B and C in Table 1. At the end of the course students had essentially shifted their perceptions to categories D, E and F. Furthermore, the students were also voicing what was categorised as more sophisticated views on learning. Their final conclusion was that 'such an epistemological framing can profoundly influence students' conceptions of science and conceptions of learning' (Linder and Marshall, 1998, p. 116).

Ryder et al. (1999) carried out a related study where they investigated how undergraduate science students developed their views about the nature of science during project work. Their investigation was focused on three different areas; the relationship between data and knowledge claims, the nature of lines of scientific enquiry, and science as a social activity. They were able to identify two key areas of development: 'the role of theory in guiding the questions which scientists investigate and the significance of critical experiments and procedures in the proof of scientific knowledge claims' (Ryder et al., 1999, p. 215).

An interesting study in this context was done by Lising and Elby (2005), who were able to show how a student's personal epistemology had a direct, causal influence on her learning. Furthermore, just as numerous tests have been developed in order to probe students' conceptions in various content areas, similar tests have been developed in order to probe students' epistemological beliefs. Two examples of such tests are the well-known Maryland

Physics Expectations Survey (Redish, Saul, and Steinberg 1998) and the Colorado Learning Attitudes about Science Survey (Adams et al. 2006).

Up to now I have been discussing studies on students' epistemologies, albeit of different kinds. However, the question of the impact of epistemology on the learning of physics can also be approached from a different perspective; by looking at how teachers' epistemological stances affect their teaching. Two examples of such studies are Linder (1992) and Hammer (1995). Linder (1992) argues that 'teacher-reflected epistemological commitments may be influencing physics teaching and its outcomes' (p. 120). In particular, he brings to the fore how a view of physics as being 'an on-going collection of mind-independent facts about objective reality' (p. 111) can be a source of conceptual difficulty among students since this view can encourage students to rote-learn facts rather than reflecting on their own understanding. Hammer (1995) takes on a different dynamic by exploring how a view of students as having epistemological beliefs can motivate a shift in teaching, from traditionally solely content-oriented towards including epistemological objectives.

2.2.4.2 Metacognition

Outside of cognitive science, metacognition is often taken to represent 'thinking about one's own thinking'; the ability to reflect upon and have control over one's own learning (see, for example, Gunstone 1991 and Georghiades 2004). Thus, metacognition is increasingly becoming considered as an important attribute of successful learning. This is particularly true for constructivists, where the metacognitive learner is typically characterised by an ability to recognize and evaluate existing ideas, and where needed, replace those ideas (Gunstone 1991). Research on metacognition within physics education is a relatively undeveloped area. Examples of studies that have been done are Linder and Marshall (1997) and Koch (2001) where metacognitive techniques for improving students' comprehensions of physics texts are developed and then evaluated. Metacognition in the context of the physics student laboratory is explored by Kung et al. (2005) and Kung and Linder (2007). They argue that it is important to consider the outcome of metacognition, not just the amount of metacognition. They further argue that whether students are encouraged to change their behaviour as a result of metacognition is dependent on the laboratory design. An excellent overview of research on metacognition, particularly in relation to science education, is given by Georghiades (2004).

2.2.4.3 Representations

Physics education research has a long tradition of research into problem-solving, in particular in regard to students' use of representations (such as mathematics, language and graphs). For example, van Heuelen (1991) has suggested that one of the keys in learning to think like a physicist is to be taught an appropriate problem-solving strategy, one that involves multiple representations. Lately, the interest in students' (and sometimes, experts) use of representations has increased. Kohl and Finkelstein (2008) have compared how expert and novice physicists use representations in their solving of physics problems. Airey and Linder (2009) depict the complex of representations, tools and activities of a discipline in terms of a 'disciplinary discourse'. They suggest that a student in order to achieve an appropriate holistic understanding of a science concept needs to become 'fluent' in a 'critical constellation' of modes of this disciplinary discourse.

Research has also been done on physics students' use of particular forms of representations, such as gestures (Sherr 2008), equations (Domert et al. 2007; Sherin 2001), graphs (Aberg-Bengtsson and Ottosson 2006; Lindwall and Lymer 2008) and language (Brookes and Etkina 2007).

2.2.5 Approaches to Teaching and Curriculum Design

Studies of students' conceptions has, together with studies of students' epistemological beliefs, influenced the development of teaching approaches within physics (van Aalst 2000). One example of such development is 'Workshop Physics' (Laws 1997), where introductory physics courses are taught without lectures. The students instead engage in, for example, discussions with teachers and peers and use computer-based laboratory tools, all aiming to create an 'active learning environment'. Another insightful example of how PER has been used to develop teaching is using a modelling perspective (Etkina et al. 2006; Hestenes 1996). Starting from a claim that construction, validation and application of scientific models is basically what scientists do, it is posited that this is also what we ought to teach our students. In other words, the focus here is on teaching students to think in a 'scientific way', rather than, for example, learning isolated concepts. By allowing the students to be included in the explicit construction of the representations used, their 'misconceptions' are argued to be indirectly challenged. Yet another example of a teaching method developed by physics education researchers is 'Physics by Inquiry' (McDermott 1991), where the teaching is embedded in the idea that 'physics should be taught as a process, not an inert body of information' (p. 306). Lately, there has been an increased interest in work to research and develop interactive simulations for the teaching of physics. An example of this is the 'Physics Educational Technology' (PhET) project that is developing simulations which seek to make visual and conceptual models used by physics experts accessible to students (Perkins et al. 2006; Wieman et al. 2008). The simulations are interactive, animated, and game-like environments, designed to engage students in active learning. Another example of such simulations are the quantum interactive learning tutorials, aimed at helping advanced undergraduate students to learn quantum mechanics, by targeting specific student difficulties and misconceptions (Singh 2008).

PER is also concerned with the evaluation of teaching. In doing so concept inventories such as the 'Force Concept Inventory' (Hestenes et al. 1992) and the 'Mechanics Baseline Test' (Hestenes and Wells 1992) have been commonly used. For example, using these tests Hake (1996) showed that the use of interactive teaching strategies enhances both students' problem-solving abilities and their conceptual understanding. Similar arguments about how students learn more from teaching that actively engages them in contrast to the kind of traditional teaching which has students typically become passive observers, are also made by, for example, McDermott (2001), Meltzer and Manivannan (2002), Crouch et al. (2004) and Crouch and Mazur (2001).

2.2.6 Learning in the Student Laboratory

Laboratory work is central to university science education, since it presents a unique opportunity to learn the essentials of scientifically based empirical activity; 'learning science by doing science' (Hofstein and Lunetta 2003). Doing laboratory work is widely considered helpful in generating an understanding of the natural world in terms of a scientific approach to enquiry (Millar et al. 1999). Learning in the student laboratory has consequently also been the subject of extensive research, as summarized in a review article by Hofstein and Lunetta (2003).

The student laboratory is a highly complex learning environment where students are expected:

- to *understand* theory (concepts, models, and laws) as described in textbooks and labsheets, or as explained during lectures;
- to *learn* concepts, models, and laws;
- to *do* various experiments, using different pieces of theory and different procedures, in order to acquire a significant experience;
- to learn to '*do again*' the same experiments, and to follow the same procedures as utilized during preceding sessions;
- to *learn* processes and approaches and be able to apply and follow them in other contexts;
- to *learn* to use scientific knowledge, think with it, as experts do, and acquire the capacity to manage during a complete investigation.

(Séré 2002, p. 625)

According to Millar et al. (1999) one of the main purposes of laboratory work is the linking of the domain of ideas to the domain of objects and observable things. Furthermore, the completion of laboratory tasks is argued to be dependent on three 'conceptual domains', namely:

> *Declarative knowledge:* knowledge and skills in practice relating to science concepts, i.e. the phenomena, laws, relationships.
> *Procedural knowledge:* knowledge and skills in practice relating to 'how to do science', i.e. the understandings underpinning the methods of scientific enquiry that the learner brings to and takes from laboratory work.
> *Communicative competence:* ability to participate in the scientific discourse community.
>
> (Rollnick et al. 2004, p. 17)

Overall, the research on laboratory work to date covers a wide variety of issues, very much in line with PER in general, such as, student conceptions, of, for example, measurements, (for example, Buffler et al. 2001; Kung 2005; Lippman Kung 2005; Volkwyn et al. 2008), student epistemology (for example, Havdala and Ashkenazi 2007; for example, Séré et al. 2001; Wickman 2004), metacognition (for example, Davidowitz and Rollnick 2003) and new approaches to teaching and the evaluation of these (for example, Allie et al. 2003; for example, Allie et al. 1997; Benckert and Pettersson 2008; Cox and Junkin 2002; Hart et al. 2000; Johnstone et al. 1998; Karelina and Etkina 2007). Research on learning in the student laboratory has further been summarized in review articles by Klainin (1988), Lazarowitz and Tamir (1994) and Hofstein and Lunetta (2003).

2.2.7 Summary of Physics Education Research Overview

My general overview of physics education research has portrayed it as a field that has progressed from relatively a-theoretical studies of students' (mis)conceptions of physics to an increased theoretical awareness and the development of theoretical frameworks modelling students learning of physics (in terms of, for example, p-prims or resources). But, not only has the field of physics education research theoretically deepened, it has also broadened, to include issues such as student epistemology, metacognition and students' and experts' use of representations.

Overall the work in PER has influenced the development of teaching approaches, curriculum design, the evaluation of teaching and work in the student laboratory.

Next I will focus on one particular aspect of the physics education research, that of studies exploring gender issues.

2.3 Physics Education Research Exploring Gender Issues[8]

2.3.1 Introduction

In their guest editorial on the future of physics education research in the American Journal of Physics, Heron and Meltzer (2005) write:

> We highlight those directions that address intellectual issues that are specific, but not necessarily unique, to the subject matter and reasoning of physics. Therefore we omit important work on investigating gender-equity issues, for example.
>
> (Heron and Meltzer 2005, p. 390)

Thus, in their discussion on the future of physics education research they give gender issues no consideration at all – claiming that such issues are not tied to physics as a discipline. Yet, within the physicist community, issues of woman under-representation (and sometime underachievement) have been intensely debated for decades. An article by Ambrosia (1940) is an early example of discussions on 'teaching physics to women'. Her main suggestions concern teaching subject matter more closely aligned with women's experiences and using students as peer-instructors in the student laboratory. Later, the issue of women/girls/gender and physics education has been intensely discussed and researched. Articles on the issue have typically taken as their starting point that the low number of women taking physics is a problem that needs to be solved. Reasons given for this can somewhat simplified be summarized as; firstly, that the science of physics would benefit either by the different perspectives brought in by women or by the increase in the sheer number of people studying physics. Secondly, along the same line, is also the argument that society as a whole is dependent on more people studying science and technology and that women is an untapped resource, in terms of numbers or new perspectives. Finally, it is argued that it is not fair that men and women today are not given the same opportunities to study physics.

In 1979 Physics Education dedicated a special issue to 'women and physics'. In this issue Ormerod et al. (1979) reports on a study of man and woman students attitudes towards physics. Taylor (1979) analyses physics textbooks for possible sexist bias. Thompson (1979) provides a statistical background to the discussion on girls and physics in terms of the number of boys and girls taking and passing physics at school and university level. Finally, Harding (1979) discusses the sex differences in examination per-

[8] A version of this section is to be published in Danielsson (forthcoming).

formance. As we will see, these articles are also quite representative for later research in the area of gender and physics education.

This part of the literature review explores studies on gender and physics education research, broadly defined. This means that not only physics education research articles have been included, but also articles from science education research dealing explicitly with physics. Journals that regularly publish articles on physics education research as well as the highest ranked science education journals and the database ERIC were searched using their online search.[9] Further, the list of annotated references on gender issues in physics/science education compiled by Mallow and Hake (2002) was also scrutinized. I also used Murphy and Whitelegg's (Murphy and Whitelegg 2006a, 2006b) review of the research on the participation of girls aged 11-16 years in science.

Whereas I, in the previous section, gave a more general overview of physics education research, this part of the review is devoted to physics education studies that share my study's focus on gender issues. Thus, the aim of this part of the review is to situate my study in relation to previous studies within physics education research which also take on a gender perspective.

After reading a large number of the abstracts fifty-seven articles were judged to be relevant to this part of the review and chosen for further analysis. I read these articles and wrote short summaries of them. The summaries were then sorted into preliminary categories. I then re-read the articles and the summaries several times focusing on how they approached the issue of gender in physics. Questions posed during the readings included: Is gender problematised or treated as a synonym for 'girls and boys'? What kind of research methodology is employed? What view of learning is expressed in the article? Is physics problematised? Finally, I was able to group the articles into five categories: 1) Comparisons of man and woman students 2) Textbooks and tests 3) Classroom practices 4) Teachers' attitudes 5) Critical perspective.

2.3.2 Summary of Findings

Broadly speaking, there are two types of publications among the examined articles: teachers sharing of experiences and ideas, and research studies. The focus of my review of gender and physics education will be on the research studies; when an article deals with teachers' sharing of ideas and experiences this is clearly indicated. The majority of research studies were found to be quantitative in nature. As discussed above, five categories were constructed empirically based on the readings of the articles: 1) Comparisons of man and woman students 2) Textbooks and tests 3) Classroom practices 4) Teachers' attitudes 5) Critical perspectives.

---

[9] The journals and the search words used are listed in Appendix A.

In the following, an overview of the articles judged to belong to each category will be given. Then, an illustrative example of one or a few studies in the category will be presented in somewhat more detail. I have chosen to structure this part of the literature review qualitatively rather than quantitatively, focusing more on exploring the occurrence of different themes and methodologies than on the number of articles applying, for example, a particular methodology.

2.3.2.1 Comparisons of Man and Woman Students

The most common way to apply a gender perspective is by a comparing of man and woman students. This can be done in terms of, performance and ability measured using grades (Hazari et al. 2007; Stewart 1998; Tai and Sadler 2001; Wee and Baaquie 1993), performance on tests (Forster 2005; McCullough 2004; Zohar 2003), relations between problem-solving performance and representational format (Meltzer 2005), interest in physics (Häussler et al. 1998; Jones and Kirk 1990; Williams et al. 2003) or attitudes towards physics (Angell et al. 2004; Reid 2003). There are also studies on the effect of high school physics preparation and affective factors on performance in introductory university physics (Hazari et al. 2007) and factors mediating the effect of gender on students misconceptions about electrical circuits (Sencar and Eryilmaz 2004). Udo et al. (2001) examines the effect of gender on students feelings of science anxiety. Woolnough and Cameron (1991) reports from an evaluation of a course in which boys' and girls' reasons for choosing physics and what kinds of assessments they preferred were compared. Another kinds of comparison is the cross-country comparison made by Menard and Uzun (1993), in which the number of women studying physics in the U.S. and in Turkey are evaluated, and the countries educational systems compared and discussed in relation to this.

One example of a study typical of this category could be that of Reid (2003). In his study boys' and girls' attitudes towards science in general and physics in particular are surveyed. Towards the end of primary school both boys and girls were found to have very positive attitudes towards science and considered it an important subject. Towards the end of the second year of secondary school, however, a significant decline in girls' positive attitudes was observed. Furthermore, girls were found to be drawn to themes perceived to have high social relevance, whereas boys were drawn to themes with high mechanical or practical relevance. The study concludes that it is important to balance the physics syllabus so that topics that 'have a natural appeal for girls as well as those preferred by boys are both included' (Reid 2003, p. 533).

Furthermore, most of the above studies are quantitative in nature, an important exception being Zohar (2003). He employs a mixed methods approach, where the statistical data are complemented by interviews with man and woman students about the possible competitiveness in the physics classroom and their valuing of understanding. An example of a qualitative study is that by Stadler et al. (2000). Based on their observations of students working together, they claim that boys and girls have different ideas about what it means to understand physics.

Common to studies in this category is that they treat gender as a stable category and focus on the differences between the genders rather than on the variations within the genders. All in all, the studies in this category construct two different kinds of physics students: man students who are interested in physics for its own sake and enjoy practical exercises and woman students who want physics taught in a way they can relate to their own lives and who have lower self-confidence, particularly in relation to practical work.

2.3.2.2 Classroom Practices

A second set of studies focus on classroom practices in different ways. These studies typically either compare how man and woman students respond to a certain form of teaching or discuss how to make the physics classroom more 'girl friendly'.

This category overlaps with the previous one, as several studies are designed to compare how man and woman students respond to different forms of teaching. Gustafsson (2005) evaluated a distance course that was altered to include cooperative work and compared man and woman students in terms of throughput and intrinsic motivation. Lorenzo et al. (2006) looked at whether the use of interactive teaching strategies in an introductory physics class could help to narrow the gender gap on the FCI. Pollock et al. (2007) did a follow-up of the Lorenzo et al. (2006) study in a different university context. Furthermore, several studies examine how man and woman students interact when working in small groups (Alexopoulou and Driver 1997; Ding and Harskamp 2006; Heller and Hollabaugh 1992; Tolmie and Howe 1993)

Most articles discussing how to make the physics classroom more 'girl friendly' are perhaps better characterised as teachers sharing of their experiences than as actual research studies. An early example of such an article is that of Pollack and Little (1973), who describe a special introductory programme offered to woman physics students, including, for example, seminars with visiting woman physicist, counselling and extensive laboratory experiences. Etkina et al. (1999) share their experiences from a physics course developed for 'at risk students' (women and minorities), where elements such as qualitative mini-labs and interactive lectures were introduced. Williams (2006) describes how he developed a unit for teaching physics on the theme 'health and beauty' aimed in particular at the less able woman students. Robertson (2006) suggests teaching practices that are aimed at ensuring a fair learning environment for both man and woman students, such as single-sex grouping in the laboratory and interactive teaching styles. Different ways of teaching physics in a more gender-inclusive way are also discussed by Parker (2002), Norby (2000) and McCullough (2007). One research study on this theme is that by Laws et al. (1999). They evaluated the efficacy of new activity-based introductory physics curricula to explore whether such an approach could have the potential to close the gap between the number of men and women studying physics. An overview of early interventions at school level designed to make the teaching more girl-friendly can be found in Taber (1991). A contemporary overview of teaching strategies that can 'help narrowing the gender gap' is provided by Lorenzo et al. (2006).

There are also projects aiming at encouraging woman physics students through, for example, early research experiences and social opportunities for them (Schneider 2001) or by high school-college interactions (Light et al. 2002). Bazley et al. (2002) discusses the Internet as a possible means for allowing woman physicists to meet each other. A comprehensive review of the reasons for under-representation of women in physics including possible ways to encourage more women to study physics, such as mentors and programs designed to introduce young women to science, is given by McCullough (2002).

Typical studies in this category could be those of Lorenzo et al. (2006) and Pollock et al. (2007). By reviewing previous research on how to reduce the gender gap Lorenzo et al. found that a common strategy suggested is that of active pedagogies. In their study, interactive methods that promote, for example, in-class interactions and that foster collaboration were applied to introductory university physics courses. The students' conceptual understanding was then tested using the Force Concept Inventory. Compared to traditionally taught courses it was found that the interactively taught courses yielded significantly increased understanding in both man and woman students and, furthermore, reduced the gender gap. Pollock et al. (2007) carried out a study similar to that of Lorenzo et al., but with a different student population. The students in the Pollock et al. study both started and ended their course with notably lower scores on the conceptual survey than did the Lorenzo et al. students, and the difference in performance between man and woman students was not reduced. They therefore conclude that interactive engagement may be necessary, but not sufficient, for reducing the gender gap.

In line with the studies in the previous category these studies typically also construct man and woman students as two different kinds of physics learners. The studies commonly have the aim of 'reducing the gender gap', whereas the studies in the previous category were instead exploring the gender gap.

2.3.2.3 Textbooks and Tests

Several of the studies examining differences between man and woman students do discuss pedagogical implications in terms of changing the context of examples etcetera to make them more gender inclusive. However, there are also studies that take the contents of textbooks and tests as their starting point, examining them from a gender perspective. Larsen (1995) investigated the inclusion of woman astronomers in astronomy textbooks. Walford (1981) and Whitely (1996) have analyzed gender balancing in physics textbooks in terms of illustrations, named scientist and gender stereotyping in illustrations and texts. Duit at el. (1992) outlines the design of a textbook with the purpose of making physics more appealing to both girls and boys. Hoffman (2002) examined boys' and girls' interests and designed an intervention project in which new teaching material was developed in order to stimulate girls' interests. Two examples of Swedish studies examining physics textbooks from a gender perspective are von Wright (1999) and Svennbeck (2004). von Wright (1999) has examined textbooks from primary and secondary school looking at, for example, their view of science, the language, the examples and how men and women are portrayed. Svennbeck (2004) focuses her examination on the 'ways of knowledge' present in secondary school textbooks and how those can be understood as gendered.

McCullough (2001) examined the contexts of the questions in the Force Concept Inventory from a gender perspective, and Forster (2005) looked at the contexts of questions in an Australian entrance examination.

A typical example of a study in this category could be that of Whitley (1996). Here seven physics textbooks were examined to determine the gender balance of the books in terms of illustrations and named scientists and possible gender stereotyping in illustrations and texts. It was found that all books included a greater number of man than woman adults and mentioned a substantially greater number of man than woman scientists. Some books, however, were found that avoided gender stereotypical illustrations, for example, by showing woman computer users.

In summary, the studies in this category in various ways critically examine how physics is presented, in terms of, for example, what contexts are chosen for examples and what people that are chosen to represent the physicist.

2.3.2.4 Teachers' Attitudes and Knowledge

Most studies focus on the students, but Zohar and Bronstein (2005) examined physics teachers' knowledge about girls' low participation in physics and their views on the issue. They found that many physics teachers are not aware of girls' low participation rate in physics, or think that the gap is smaller than it actually is. Furthermore, about two-thirds of the interviewed teachers did not see girls' low participation in physics as a problem that requires any action. They conclude that 'the data show that many of the reasons teachers gave are based on powerful gender stereotypes, expressing the view the men and women are born with different intellectual capabilities and that they are destined to different types of social and professional roles' (Zohar and Bronstein 2005, p. 73).

2.3.2.5 Critical Perspectives in Physics Education Research

Common to the studies presented previously, with the possible exception of Zohar and Bronshtein (2005), is that they do not critically examine the meanings of science, but rather see physics as something relatively fixed. Critical perspectives on gender and the learning of physics are rare, but one example of such a study is that by Carlone (2004). She takes as her starting point recent literature on girls and school science arguing that in order to engage girls in science, educational activities need to promote broader meanings of science and scientist. Her ethnographic study then examines the meanings of science and the kinds of science identities produced by students in a reform-based physics classroom (Active Physics). She found that the girls both resisted and accepted the active science learner identity. Carlone argues that when the girls resisted the active learner identity, it was because this identity threatened their highly valued 'good student identities', that is, their perception of what it meant to be a good student.

Despite the interactive and therefore presumably girl-friendly Active Physics curriculum, Carlone demonstrates how the curriculum enacted in the classroom she studied promoted meanings of science as difficult and hierarchical. This is well in line with Pollock et al.'s (2007) conclusion that interactive teaching approaches are not sufficient to promote woman physics learners, but that what it in fact needs to be studied is how a particular interactive teaching approach is enacted by faculty and students. Summarizing, Carlone writes:

> The difficult and hierarchical nature of the enacted Active Physics curriculum implied science identities (e.g., someone who is "naturally" smart, has "raw talent", and is man) that were alienating, inaccessible, and/or uninteresting for girls. At the same time, these meanings did not challenge girls' taken-for-granted assumptions about who is "good" at science. Thus, most girls (even the successful ones) did not actively resist these celebrated science identities unless they perceived the practices as threatening to their grades or their "good student" identities.
>
> (Carlone 2004, p. 405)

In comparison with the studies presented previously, Carlone views both gender and the learning of physics from quite a different perspective. First, she looks at learning as identity formation, rather than acquisition of knowledge, and she is gives agency to the learners, in that they are actively resisting and accepting different physics learner identities. Second, she critically examines a physics curriculum and the identities that are made possible by this curriculum.

2.3.2.6 Another View of Physics Education and Gender

The studies presented in section 2.3 could be characterised as belonging to physics education research and science education research. Issues of gender and physics education have also interested researchers outside the immediate physics education researchers' community. In her anthropological study of high energy physicists Traweek (1988), among other things, explored how physicists learn, produce and re-produce the culture of physics throughout their career, from undergraduate student to professors, and also how this process in various ways is influenced by gender.

A seminal study of gender and the learning of physics is also that of Thomas (1990), where she compares and contrasts the experience of being the minority gender in undergraduate physics and literature studies respectively. Like Traweek (1988) Thomas is also interested in the culture of academic disciplines and students' identity formation in relation to this culture. Nespor (1994) also addresses the issues of gender and physics learning in his actor-network inspired study of undergraduate physics and management students. More recently the anthropologist Hasse (1998, 2000, 2002a, 2002b, 2006) has explored issues of gender and physics learning. In particular Hasse is interested in what activities contribute to inclusion and exclusion in physicist communities.

2.4 Gender and Science and Technology Education

2.4.1 A Brief Historical Perspective

The participation of women in science is a highly debated area and commonly the focus has been on how to attract more woman students. There have been, and still are, however, several different ways to view 'women and science', or 'gender and science' for that matter. The most important shift was probably what Harding characterised as 'from the women question in science to the science question in feminism' (Harding 1986). In other words, instead of viewing women as the 'problem' that needs to be fixed, for example, that they need to learn to think more like men, it is science that is viewed as the 'problem', science itself needs to be critically examined.

Berner (2003) and Johnstone and Dunne (1996) examine the assumptions about gender and science that have underpinned discussions about a more inclusive science teaching. Johnstone and Dunne describe one type of research as centrally concerned with documenting differences in achievement or participation, sometimes seeking explanations for these differences in biology. Berner (2003) characterises this research as 'sex-roles research'. Another strand of research, described by Johnstone and Dunne (1996), seeks social explanations for gender differences, such as the effect of parental influence. What both the 'biological' and the 'socialisation' perspectives have in common is an epistemological view that their findings reveal 'facts' about boys, girls, and science. Conclusions are of the type that girls, for example, prefer a certain kind of learning environment. From this perspective 'a change in the situation, then, requires either girls to have experiences that compensate for their deficiencies or for the school learning environment to be altered to compensate for the learning styles of girls' (Johnstone and Dunne, 1996, p. 58). Further,

> What must be recognised here is that the oppositions that are constructed, within both the research and the interventions which are developed from it, are constitutive of gender. They produce and reproduce the categories that they are assuming to describe. Ironically, in this production, the relationship that the research is seeking to challenge – the dominance of the masculine over the feminine – is reproduced through these oppositions.
>
> (Johnstone and Dunne 1996, p. 59)

What Johnstone and Dunne (1996) are arguing for is research that engages with the dynamics of gender construction, that looks at how the dualistic gender relation is produced and reproduced in social practices, practices in which the said research is a part.

Berner (2003) describes how contemporary research more and more has turned away from the previous, often very passive view of woman students as 'victims', either of biology or of socialisation, and now instead focuses on their conscious choices. Gender is in this view seen as a question of choice and performance, rather than biological inherent behaviours or socialised norms. Here, focus has more and more moved towards looking at variations within the genders, their dynamics and diversity, not viewing them as a simple dualism.

2.4.2 Some Contemporary Approaches

Contemporary research on science and technology education has seen a shift described by Harding (2005) as:

> ...the emergence of a critical focus on the masculinized culture of science and science education, and on how 'doing science' is a way of constituting certain kinds of social identity.
>
> (Harding 2005, p. 244)

One strand of this research is dealing with girls' engagement in (elementary) school science, examining how girls construct science student identities. An example of this is Brickhouse et al. (2000) whose aim is to understand the variety of ways girls engage in science and how this engagement is shaped by their views of themselves. Four student stories are presented and analysed, and the diversity in these stories is striking – in sharp contrast to how much earlier research has considered girls only in comparison to boys, thereby exaggerating the differences between the groups and not being sensitive to the diversity within the groups. However, while Brickhouse et al. (2000) were able to identify a wide variety of girls' ways of engaging in science they could also conclude that the girls that took on more traditional gendered identities (doing 'good-girl' student identities) were the most likely to be encouraged by their teachers. A related study is that of Brickhouse and Potter (2001) who examine the 'scientific identity formation' of two young African-American women in an urban vocational high school, describing how experiences of marginalisation can make membership in school science communities both impossible and undesirable. Yet another example is Ford et al. (2006) who have studied elementary school girls' reading of science books at school and at home in the light of how young girls often construct strong identities as readers and writers, but find it difficult to construct scientific identities. They conclude that 'narratives may be an important means of engaging girls in science, as the girls in the study seemed most interested in scientific information that was embedded in an engaging story' (p. 286). Tan and Calabrese Barton (2007) made an ethnographic study of two minority girls' participation in science, exploring how these two girls engage in science. They conclude:

> The purposeful authoring of novel identities-in-practice by both girls also made manifest to us their sense of agency and interest in school science. … In using identity as a lens to understanding how minority girls participate in science, we have a deeper understanding of how they display agency in working to succeed in science.
>
> (Tan and Calabrese Barton 2007, p. 69)

A common theoretical ground in this type of research is a conceptualisation of learning as identity formation, giving rise to an interest in how students engagement in science is related to who they are and who they want to be (Brickhouse 2001). Identities are here understood as multiple and fluid and an aim is often to describe the diverse and varying ways that students engage in science, in and out of school, some of which may not be traditionally understood as 'science learning'. Further, there are accounts of how students both adopt and resist science student identities (see also Carlone 2004). It should be noted though that the focus is exclusively on girls and their participation in science; boys' participation in science is seldom problematised.

Related to this research on girls' participation in science is also a (feminist) discussion on the purposes of school science education, often concerned with how school science works with too narrow conceptualisations of science, related closely to training (traditional) scientists. Working from the premise that 'science education programmes should aim to educate students *about* science, not train them to *be* scientists' (Gilbert 2001, p. 300), Gilbert develops an argument about how schools ought to develop students' critical literacy with respect to science. In other words, to help them develop the same relationship to science that an art critic might have to art. Similar arguments are also presented by Letts (2001) and Kumashiro (2001), who both advocate a school science aimed at developing critical scientific literacy. Letts writes:

> In science class, not only can a variety of texts be examined and deconstructed, but the activities and experiments that constitute the 'doing' of science can also be interrogated as part of the science learning.
>
> (Letts 2001, p. 270)

A related approach is taken by Sjøberg (2000) who discusses how science can be understood as a subculture, with certain characteristics and advantages, but also clear delimitations. By bringing this to the fore in the teaching of science Sjøberg (2000) claims that the students can be made aware that the subculture of science is one that people can move in and out of; that being a scientist is not a person's whole identity. Brickhouse (2001), on the other hand, discusses how science classes could be a place where students 'could take up identities like environmentalist, feminist, and smart healthcare consumer' (p. 289) thereby connecting the school science to communities relevant to the students outside of school.

Another strand of research that also critically examines the culture of science and technology deals with women's participation in engineering, often from the perspective of organisation research. Examples of such research are Kvande (1999) and Jorgensen (2002), who both examine the available identities for women within engineering, with a particular focus on the reconciliation of feminine subjectivities with a man-dominated organisation. They both describe a number of different positionings available to woman engineers, with a particular focus on how their self-positionings affirm or challenge traditional gender stereotypes. An example of a positioning described by Jorgensen is 'positioning the self as career identified'; 'the research participants cast themselves not only as academically proficient in math and science but also as singularly called to the profession' (Jorgensen 2002, p. 363). Similar research focuses are also found in Henwood (1996, 1998), Phipps (2002), Stoyner (2002), Gill et al. (2008) and Walker (2001) who all in various ways explore the discursive production of gender among woman engineers; how woman engineers adopt and/or resist traditional gender roles. A recurring theme in this research is how the woman engineers similarly construct themselves as different from the man engineers and also different from other women, struggling with contradictory discourses about gender differences as natural, yet vulnerable to change. For example, by constructing themselves as 'different' from other women, by a counter-identifying with traditional femininity, the woman engineers in Walker's (2001) study are able to constitute engineering identities, but without challenging dominant gender roles. A similar study, but in the context of undergraduate science programs is that of Erwin and Maurutto (1998) who found that:

> [The woman science students] are constructed as both the 'exception' and the exceptional – the smart girl, the smart black girl – the woman who is different from/better than other girls by virtue of talent and desire for a 'male' interest and vocation. There are few alternative discourses in which they can make sense of themselves, of their desires, conflicts and anxieties, or imagine a different way of being.
>
> (Erwin and Maurutto 1998, p. 65-66)

Common for these studies is a focus on the 'doing of gender' rather than the 'doing of engineer/scientist'; the discourse of engineering/science is seldom explored beyond being seen as an expression of hegemonic masculinity. As with the previously described research on girls' engagement in science, the focus on identity constitution within engineering has also almost exclusively focused on women.

2.5 Situating of my Study

In the following section I discuss the previously presented research in order to demonstrate how my study is situated in relation to, in particular, physics education research dealing with gender issues, but also in relation to PER in general and in relation to research on gender and science and technology education.

The majority of studies on gender and physics education were found to be quantitative and dealing with students in primary and secondary education or introductory university courses. Qualitative studies as well as studies dealing with students majoring in physics and physics teachers are thus lacking. Furthermore, most studies were also found to be relatively a-theoretical in their application of a gender perspective, often taking gender as a synonym for biological sex. This puts a focus on the differences between men and women, rather than the differences between men and between women. Further, most studies construct man and woman physics learners as two distinctly different groups, even though some studies do acknowledge the variations within the genders (see, for example, Zohar 2003). Implicitly or explicitly the view of learning in the studies with a gender perspective is one of learning as the acquisition of knowledge about the physics subject matter, where the students are seen as relatively passive recipients of knowledge (with the exception of Carlone 2004). This is well in line with how, traditionally, research about the teaching and learning of physics has focused extensively on understanding and improving learning outcomes in terms of students' difficulties with physics-content knowledge. An implicit assumption in such research is that the student who performs well in physics is striving to be a member of the physicist community. Seymour and Hewitt (1997), however, in their social anthropological study 'Talking about leaving' clearly demonstrate that it is not necessarily so straight forward. This is particularly so for students from non-traditional groups (for example, woman physics students), since many of these, despite excellent achievements, choose to leave their science studies because they do not feel comfortable with the educational environments they find themselves in at science departments. The physics education which strives to include students in a future physicist community can thus only function if the students see this as something that is compatible with who they are and who they want to be.

The physics education articles with a gender perspective which focused on classroom practices also typically construct men and women as two relatively homogenous groups; many of them could be characterised as teachers' sharing of experiences of how to make the physics classroom more 'women friendly'. While many of the strategies described most certainly can have positive consequences for both men and women, I agree with Sowell (2004) that:

> By promoting 'female-specific intervention programs' the complexities of gender are not fully explored; in fact, they are greatly reduced. The underlying assumption is that all women express the same form of femininity and that all would benefit from the same type of 'female appropriate teaching strategies'. I would argue that such pedagogical approaches work to sustain existing gender inequalities by promoting the naturalness of what we consider male and female.
>
> (Sowell 2004, p. 58)

Furthermore, principally as a result of cognitivist and individual constructivist perspectives, learning is, within physics education research, most often understood as an individual endeavour, with a focus on the individual student's sense-making or their use of 'cognitive resources' (see, for example, Thacker 2003; Hammer et al. 2004; Redish 2004). Consequently, when such research employs a gender perspective it should come as no surprise that it tends to see gender as a characteristic of the individual students, but without relating the students' learning to the gendered characteristics of physics as a discipline. In order to more fully understand the learning of physics I would therefore argue that a shift is needed from a focus on the individual's performance to 'how students engage in science and how this is related to who they are and who they want to be' (Brickhouse, 2001, p. 286). This would then in turn also allow for a shift from asking questions about how men and women *do* physics to how women and men are *done* in physics – how they constitute gendered identities along and against the gendered norms of physics. Such a shift is recognisable in the broader realm of science education research, where, in particular, studies examining girls' engagement in school science have emerged over the last ten years or so. This research clearly demonstrates the need to take into account a diversity of ways of engaging (and not engaging) in science; to move beyond the rigid dualism of men/women in order to capture the diversity of identities – masculine and feminine – available within a science community of practice. As Brickhouse et al. (2000) concludes:

> …when teaching girls science and trying to explain why it is they are or not doing well in science, we need to know more than that they are girls. We need to know what kind of girls they are.
>
> (Brickhouse et al. 2000, p. 457)

Similarly we need to know 'what kinds of women and men' are participating in physics – and how they do various masculinities and femininities in relation to the gendered norms of the physicist community of practice. This is in sharp contrast to how physics education research largely has treated gender as an 'independent variable'; as something static, often used to sort students into categories of men and women. However, studies of/from the related disciplines of engineering and 'science' provides important insights, empirical as well as methodological, into how students and professionals 'do gender' in these disciplines. Nevertheless, there is a certain tendency in these studies to provide a detailed analysis of the students'/professionals' doing of gender at the expense of an equivalently detailed understanding of their doing of engineer/scientist. To some extent this lack in detail of the students'/professionals' disciplinary doing can probably be understood as a result of the broad areas under study; engineering or science. Consequently, I would argue that much of the research on women's participation in engineering/science could apply to any kind of man-dominated work organisation. Of course, this does not make it any less interesting or important (possibly the opposite), but I would argue that it needs to be complemented with more detailed analyses of women's and men's participation in specific sub-disciplines (such as astrophysics) or particular activities (such as fieldwork). This is my argument for taking the particular activity of doing laboratory work in physics as the starting point of my research, thereby allowing for a more detailed analysis of the 'doing of physics'.

As previously pointed out, earlier studies have commonly focused on one of the two 'doings'; either gender or science is in some sense 'black-boxed'. For example, in Nespor's (1994) detailed analysis of physics students' activities, their trajectories, their production of the discipline of physics there is an underlying implicit assumption that even though physics most clearly is something they do, 'men' and 'women' is something they just are:

> The women in the program, then, were only 'partially enrolled' in the physics actor-network. Overwhelmed numerically in the program, marginalized in the study groups, socially isolated, women might succeed in finishing the pro-

> gram with good grades, but seemingly at the price of their identification with the field.
>
> (Nespor 1994, p. 44)

Thus, in Nespor's (1994) study the identification with physics is changeable, the gender identity is not. In summary, my argument here is that in order to gain a deepened understanding of students' participation in physics, and in particular the gendering of this participation, we need a theoretical foundation that allows for a simultaneous analysis of the doing of physics and the doing of gender. I argue that a combination of situated learning theory and post-structural gender theory has the potential of providing such a foundation, which I will develop in Chapters 3 and 4.

PART II

Conceptual Framework

Introduction to Part II

In this, the second part of my thesis, I present the conceptual framework of my research. Chapter 3 introduces the broader theoretical staging, which in Chapter 4, is focused into my theoretical framework. In Chapter 5 the analytical tools are presented. Chapter 6 includes a discussion of methodological considerations. The contents of Chapters 3 to 6 and their interrelations are further unpacked in the following.

In my review of physics education research in the previous chapter I showed how this research has been dominated by individualistic perspectives of learning, often focused on the individuals' acquisition of content matter. Further, I argued that the understanding of learning as an individual endeavour could at least in part explain why, when physics education research applies a gender perspective, gender is generally seen as a trait of the individual rather than something 'done' in a social context. My argument is that in order to more fully understand students' learning of physics the studies of how individuals acquire knowledge and skills need to be complemented with studies drawing on broader conceptualisations of learning. Brickhouse (2001) writes:

> ...in order to understand learning in science, we need to know much more than whether the students have acquired particular scientific understandings. We need to know how students engage in science and how this is related to who they are and who they want to be.
>
> (Brickhouse 2001, p. 286)

Such a broadened understanding of learning also allows for a corresponding broadening of the understanding of 'gender'; viewing both 'learning' and 'gender' as aspects of identity constitution. As stated in the theoretical purpose in section 1.1, an important part of this thesis is to contribute to such a broadening within the realm of physics education research; 'to formulate a framework that allows for the exploration of physics students' gendered identity constitution'. In the following three chapters I present the theoretical and analytical framing I have developed for this purpose – the conceptual framework of my study. In Chapter 3 I give a general overview of the theoretical staging of my research; how central themes such as 'gender' and 'learning' are conceptualised within post-structural gender theory and situated learning theory respectively, and also how these theories can be understood as sharing important common grounds. In Chapter 4 I introduce my more focused theoretical framework, a tool of inquiry I have developed in close association with my empirical material. In Chapter 5 I present the concrete analytical tools I used to approach the empirical material, which consists of transcripts of semi-structured interviews with university physics students. In total I conducted 22 interviews. In the first round of interviews in the spring of 2005 I interviewed thirteen undergraduate students; in the second round in 2007 I interviewed nine graduate students.[10] More information about the interviews, as well as other methodological considerations, can be found in Chapter 6.

All in all, the conceptual framework of my research, 'the system of concepts, assumptions, expectations, beliefs and theories that supports and informs [my] research' (Maxwell 1996, p. 25) consists of three different, but interrelated, parts: The broader theoretical staging of situated learning theory and post-structural gender theory, my theoretical framework, and the analytical tools. The aim of this conceptual framework is to guide the research process, to give it order and coherence (Cohen et al. 2000), but the development and sharpening of the conceptual framework has also been an important part of the research process. From this perspective, the order of Chapters 3 to 5 also represents the chronological development of my conceptual framework.

As recommended for qualitative research (see, for example, Ely 1991) I started my research with a very broad research interest, rather than a focused research question. Coming from a background in experimental physics, the actual doing of physics, the laboratory work interested me, and that was what I initially decided to focus on. To explore the learning in the laboratory I conducted pilot interviews with three engineering students. Following one of the interviews I had a more informal conversation with one of the interviewees about, amongst other things, her background and how it had affected her choice of study and study habits. For me this conversation brought to the fore how very different identities could be constituted by students within the same studying environment, and I started to investigate situated learning theory as a way of exploring this identity constitution. This broader exploration of situated learning theory later grew into the discussion of this theory found in Chapter 3. However, parallel to my interest in laboratory work I had also started my research with a sense that 'gender' could be important for understanding the learning of physics, if for no other reason than because of the dominance of men in physics. Focusing my reading on situated learning theory I became more and more inspired by Wenger's (1998) notion of learning as the joining of a community of practice and also came

[10] With 'undergraduate students' I mean students enrolled in the Master of Science programme prior to the Master's research project. With 'graduate students' I mean students currently enrolled either in Master's or PhD projects.

across Paechter's work on masculinities and femininities as communities of practice. Together these parallel understandings of the 'doing of physics' as participating in a community of practice (Wenger) and the 'doing of gender' as participating in communities of masculine and feminine practice (Paechter) became the foundation of my theoretical framework. This theoretical framework was further developed in close association with the analysis of my interviews with undergraduate physics students, and the final version is presented in Chapter 4. However, despite being developed in close association with the empirical material I found it increasingly difficult to use the theoretical framework directly 'on' the empirical material. This need for concrete analytical tools to bridge the gap between the theoretical framework and the empirical material became particularly pertinent as the framework was challenged with new empirical material; the transcripts from my interviews with graduate students in physics. At this stage I therefore started exploring the possibilities drawing on different analytical tools and finally settled for Discourse models and positionings, which are presented in Chapter 5. Having complemented my conceptual framework with these analytical tools I (re)analyzed the entire empirical material, and it is this entire analysis that is presented in Part III of the thesis.

CHAPTER 3
Theoretical Staging

Structure of the Chapter

In this chapter I introduce the theoretical staging of my thesis. Firstly, I will discuss situated learning theory, with a particular focus on the theory's relation to issues of gender and power. Secondly, I will move on to discussing post-structural gender theory, in particular the notion of masculinities and femininities as communities of practice.

3.1 Situated Learning Theory

When Lave (1993) characterises situated learning theory she does this by contrasting it with cognitive theories of learning:

> Traditionally, learning researchers have studied learning as if it were a process contained in the mind of the learner and have ignored the lived-in world. … Theories of situated activity do not separate action, thought, feeling, and value and their collective, cultural-historical forms of located, interested, conflictual, meaningful activity. Traditional cognitive theory is 'distanced from experience' and divides the learning mind from the world.
>
> (Lave 1993, p. 7)

In other words, a key aspect of situated learning theory is how it does not separate the known from the knower, thinking from doing or the emotional from the rational – it is in fact working against such dichotomies (Lave 1988). Further, Lave (1996) calls for an increased focus on 'learning' within educational research; she argues that learning ought to be understood as the fundamental phenomenon, which teaching may or may not be a part of.

In situated learning theory knowledge is thus understood as being 'situated', meaning that it is a product of the activity, context and culture within which the knowledge is developed and used (Lave and Wenger 1991). The accompanying social structures of a given practice are seen as providing, and thus defining, the possibilities for learning. In other words, learning itself is viewed as contextually bound activity; the context does not only shape how things are learnt but also what can be learnt.

Central to situated learning theory is the concept of 'communities of practice'. In its simplest form a community of practice can be understood as a group of people engaged in some kind a shared practice in pursuit of some kind of common goal (an example could be a football team). In developing his conceptualisation of a community of practice Wenger (1998) makes use of two different, but interrelated, perspectives, those of practice and identity. In short, the practice perspective describes the community as a whole, in terms of, for example, how its boundaries towards other communities are set. The identity perspective is concerned with how the individual participants relate to the community of practice.[11]

Another key concept in situated learning theory is that of 'legitimate peripheral participation', through which newcomers to the community develop their expertise and eventually become integrated members of the community (Lave and Wenger 1991). In essence Lave and Wenger (1991) constituted a model of learning that brought participation to the fore; participation is here portrayed as a way of learning, 'of both absorbing and being absorbed in the "culture of practice"' (p. 95). In this model, newcomers first participate in activities that are not central to the practice and then move on to increasingly complex and important activities. During the period of legitimate peripherality the apprentices gradually get involved in the practice of the community and get to know the community of practice in terms of, for example, 'who is involved; what they do; what everyday life is like; how masters talk, walk, work, and generally conduct their lives; how people who are not part of the community of practice interact with it; what other learners are doing; and what learners need to learn to become full practitioners' (p. 95).

Lave and Wenger based their model on studies of five apprenticeship scenarios. Wenger (1998) later developed the model (in particular, the concept communities of practice) further, extending it to organisations. In summary, learning – from the perspective of the community of practice – has been described by Wenger (1998) as involving the following processes:

- *Evolving forms of mutual engagement*: discovering how to engage, what helps and what hinders; developing mutual relationships; defining identities, establishing who is who, who is good at what, who knows what, who is easy or hard to get along with.
- *Understanding and tuning their enterprise:* aligning their engagement with it, and learning to become and hold each other account-

[11] The concept of community of practice will be further developed in the next chapter as I construct my theoretical framework.

able to it; struggling to define the enterprise and reconciling conflicting interpretations of what the enterprise is about.

- *Developing their repertoires, styles, and discourses*: renegotiating the meaning of various elements; producing and adopting tools, artifacts, representations; recording and recalling events; inventing new terms and redefining or abandoning old ones; telling and retelling stories; creating and breaking routines.

(Wenger 1998, p. 95)

In this way the newcomers not only develop their expertise in the practice itself, but also develop an understanding of the surrounding culture. Thereby, situated learning theory focuses strongly on the individual, but as an individual-in-the-world; a member of a socio-cultural community. Lave and Wenger (1991) describe this as follows:

> As an aspect of social practice, learning involves the whole person; it implies not only a relation to specific activities, but a relation to social communities – it implies becoming a full participant, a member, a kind of person. ... To ignore this aspect of learning is to overlook the fact that learning involves the construction of identities.
>
> (Lave and Wenger 1991, p. 53)

Learning can thus be understood as 'a process of coming to be, of forging identities in activity in the world' (Lave 1992, p. 3), in short; an identity transformation. In the case of my research, not only do the students learn to do physics (in the sense of acquiring knowledge and skills), they also learn to become physicists. In the context of learning in laboratory contexts students can thus be understood as constituting identities as physics students (and possibly physicists) and also negotiating the norms of what it can mean to 'do physics'; they are together constituting a localised physics student community – partly in response to the broader physicist community of practice. However, it is also important to recognize that the students themselves are producing and re-producing norms for being a physicist as they 'educate' one another in the appropriate ways of 'doing physics'. Sfard (1998) eloquently summarises this dual process of learning as participation and negotiation:

> ...learning a subject is now conceived of as a process of becoming a member of a certain community. This entails, above all, the ability to communicate in the language of this community and act according to its particular norms. The norms themselves are to be negotiated in the process of consolidating the community.
>
> (Sfard 1998, p. 6)

From this perspective learning is not just an acquisition of memories, skills and routines; it is about forming an identity. By participating in the practice in our particular way we also contribute to making the practice what it is, thus, 'our experience and our membership inform each other, pull each other, transform each other' (Wenger 1998, p. 96).

Thus, learning involves the whole person; it implies not only a relation to specific activities (Lave and Wenger 1991). What you are becoming crucially and fundamentally shapes what you 'know' and Lave (1996) therefore argues that knowledge (or 'knowing' rather) ought to be understood as something in the doing, not something you have. Consequently, an activity, like the doing of laboratory work in physics, cannot be understood as an isolated entity:

> Learning is not merely a matter of acquiring knowledge, it is a matter of deciding what kind of person you are and want to be and engaging in those activities that make one part of the relevant communities.
>
> (Brickhouse 2001, p. 286)

It can be noted, though, that situated learning theory's modelling of 'learning' is far from uncontroversial. When, for example, Nespor (1994) describes a similar process, he abandons the concept of learning altogether, a decision based on how learning is so strongly associated with an individual, internal process. Instead he talks about 'the pathways and trajectories that entangle [the learners] in the discipline's representational productions of space-time' (Nespor 1994, p. 53).

In this overview of situated learning theory I have chosen to present it as a relatively consistent theory, which, needless to say, is a significant simplification. Being applied in contexts as different as sociolinguistics and organisational theory (where in particular Wenger's development of the concept of communities of practice has gained enormous popularity) the variation in how situated learning theory has been interpreted and applied is considerable.

Finally, it can also be noted that this overview of situated learning theory has deliberately focused very little on Wenger's development of the theory, and in particular the concept 'community of practice', as this is extensively addressed when I develop my theoretical framework in Chapter 4.

3.1.1 Situated Learning Theory and Gender

As indicated earlier situated learning theory is a theory of learning employed within several different disciplines, such as education, organisational studies and sociolinguistics. The applications of situated learning theory within organizational studies have been criticized by Salminen-Karlsson (2006) for being largely gender blind and she strongly advocates for the inclusion of a gender perspective into such studies:

> Given that communities of practice really are important to their members' identity building, they represent important sites for doing gender and learning to do gender. As Lave and Wenger propose, the ultimate result of learning is the acquisition of an identity as a valuable member of a community. But it should be added that this also means becoming a gendered member of a community, behaving in a way that the community finds gender-appropriate.
>
> (Salminen-Karlsson 2006, p. 42)

Within science education research there are most certainly also gender-blind applications of the theory (see, for example, Sweeney and Paradis 2004), but here the situated learning theory has received particular interest among researchers interested in the gendering of science learning (see, for example, Tan and Calabrese Barton 2007). This interest can be traced back to an article where Brickhouse (2001) compellingly argues that situated learning is the theory of learning that has most to offer when examining science learning from a feminist point of view. She bases her argument on the many epistemological and historical similarities between situated learning theory and feminism. Drawing on the work of Lave (1988) she exemplifies how situated learning theory, as rooted in critical anthropology, has been concerned with how the 'natives' under study have been portrayed as the opposite of the ideal rational scientific man. The 'natives' have thus been described as irrational and emotional, more closely associated with nature than with culture. Not only did this portrayal limit the anthropologists' understandings of the natives, it also justified treating them as less than human. This definition of the 'natives' in opposition to the rational scientist is, Brickhouse (2001) argues, closely related to how women in Western cultures have been described and their subordinate position justified. Thus, feminists and situated learning theorists do have overlapping political projects. This also includes their view that science and rationality are best understood as local practices.

Furthermore, situated learning theory's view of learning as an identity transformation enables a conceptualisation of the gendered nature of learning that stretches far beyond how gender, as argued by Glasser and Smith III (2008), within education research all too often has been used as a synonym for 'sex', which limits the use of the notion to a way of categorising learners into men and women. Situated learning theory calls for a fundamental reconsideration of what questions we ask about learning; rather than focusing on whether a student has acquired a particular scientific understanding it focuses on 'how students engage in science and how this is related to who they are and who they want to be' (Brickhouse 2001, p. 286):

> ...a girl who sees herself as someone who needs credible explanations for how the world works and aspires to understand this scientifically, may engage in science differently than a girl who wants to be an 'A' student but does not aspire to know more about science than is required of her.
>
> (Brickhouse 2001, p. 286)

In this view gender and science are thus 'done' simultaneously; the doing of science and the doing of gender are inseparable, and Brickhouse (2001) writes:

> The idea of learning as transformation of identity-in-practice provides a way of thinking about learning that is gendered, but does not regard gender as a stable, uniform, single attribute. We are not born with gender. We do gender.
> (Brickhouse 2001, p. 290)

For example, Salminen-Karlsson (2006) concretises the entanglement of doing gender and participating in a community of practice in the following way:

> When individuals accept and integrate the meaning of the practice they belong to, their identity changes. But the prerequisites of identity-building are not just a gift from a community of practice: community membership also sets limits on an individual's identity building. ... When workplaces, or communities of practice, have implicit expectations as to how femininity and masculinity should be 'done' these expectations join other messages as part of the material for constructing the individual's work identity.
> (Salminen-Karlsson 2006, p. 34-35)

Furthermore, viewing learning as an identity transformation calls for a focus on the individual, but not on the individual as an isolated entity, but as an individual that is situated within social structures, with no division between the two:

> It accounts for the importance of both individual agency as well as societal structures that constrain individual possibilities, both of which are necessary for any adequate understanding of gender relations. We know that individuals are not free to be anyone they wish. ... However, we also know that society does not totally define a person.
> (Brickhouse 2001, p. 286)

In summary, situated learning theory makes social categories, such as gender, central for understanding student-learning. Consequently, Brickhouse (2001) has argued for situated learning as being a powerful framework when seeking to understand the gendered experience of learning science.

Following Brickhouse's suggestion situated learning theory has been used in numerous studies researching the gendered experience of learning. Good examples are Brickhouse and Potter (2001), Case and Jawitz (2004), and Du (2006). Brickhouse and Potter (2001) studied how young women in an urban context form scientific identities and how this identity formation is affected by their experience of marginalisation due to gender or ethnicity. Case and Jawitz (2004) studied how issues of race and gender affect South African engineering students' experiences of vacation work. Du (2006) studied engineering students gendered identity formation in a problem-based learning environment. However, in this context a mentioning of Hildebrand's (2001) feminist critique of situated learning theory is needed. Hildebrand takes as her staring point the concept 'legitimate peripheral participant' and argues that learning through apprenticeship pacifies the learner because it is really just a copying of the 'master' without any possibility of critiquing the practice of the community. However, I would argue that Hildebrand's critique is partly due to a somewhat limited view of Lave and Wenger's work; that learning is viewed as the inclusion of a newcomer into a community of practice is not necessarily the same as that '[a]ll of their practices and discourses are assumed as exemplary and beyond critique' (Hildebrand 2001, p. 8). Or, at least, that this does not necessarily have to be the case.

3.1.2 Situated Learning Theory and Issues of Power

As pointed out by Salminen-Karlsson (2007), introducing the concept of gender into a situated learning approach also requires the introduction of a concept of power. Contu and Willmott (2003) are, in their critical examination of Lave and Wenger's (1993) original work as well as popularisations thereof (in particular Brown and Duguid 1991), able to show how issues of power play a central role in the theoretical outlining of Lave and Wenger's original work, but how popularised versions of situated learning theory as well as Lave and Wenger's own empirical examples ignore or suppress the 'understanding that learning processes are integral to the exercise of power or control' (Lave and Wenger 1991, p. 284).

Consequently, situated learning theory is often criticised for its lack of consideration of issues of power. For example, the sociolinguist Davies (2005) argues that the 'communities of practice perspective' is in need of further development in terms of a communities' internal hierarchy. Furthermore, Davies (2005) argues that 'implicit with the current concept of communities of practice seems to be the idea that individuals have some degree of choice in the extent of their membership in a community' (Davies 2005, p. 576), a view she goes on to argue underplays the importance of legitimacy in order to gain access to the practice. Thus, mechanisms within communities of practice for gaining and maintaining membership need further investigation. An empirical exploration of the issues raised by Davies (2005) in relation to hierarchy, power and status in communities of practice can be found in Moore (2006).

Salminen-Karlsson (2006) is also critical about how theoretical work about communities of practice generally has paid very little attention to external influences on the community. She argues that Wenger (1998) 'does not reflect on the extent to which general ideologies, economic conditions and power structure both in the society and in the surrounding organisation affect the practice of a community' (Salminen-Karlsson 2006, p. 34). According to Salminen-Karlsson (2006) Wenger does not only overlook external influences on the community as a whole, but also overlooks how individual members are affected by conceptions of, for example, gender, thereby treating the individuals more or less as free agents. In principle, Salminen-Karlsson argues for the inclusion of a Foucaldian view of power into the communities of practice perspective in order to analyse how power relations are embedded in the productions of meanings, identities and knowledge. She points out how 'power is not only acquired in the course of the personal participation trajectory in the community, but also accorded externally, for example, by the direct influence of the institutional framework of the community of practice' (Salminen-Karlsson 2006, p. 37). Furthermore, the power relations in a community are also affected by the societal power systems (related to, for example, gender) that the members bring with them. In Foucauldian terms this would be understood as the intersection of different discourses, and 'that these discourses define the different possibilities for individuals to shape both their identities and their daily working environment' (Salminen-Karlsson 2006, p. 37). Salminen-Karlsson (2006) elaborates:

> The Foucauldian concept of power can be seen as particularly well suited to the analysis of power in communities of practice in combination with the doing gender approach, as it views power as something that is 'done' in different ways, especially in discursive interaction.
>
> (Salminen-Karlsson 2006, p. 37)

In this view power is not seen as something only some people exercise over others, but rather as a 'continuous play' (Salminen-Karlsson 2006). As summarised by Paechter (2006c): 'Power, for Foucault, is also not conceptualised as descending from above, but rather, operates in a capillary fashion, throughout society' (p. 16). Salminen-Karlsson (2006) exemplifies how the old-timers of a community may well enjoy a certain vested power in that the practice has been formed according to their needs, but as newcomers are also necessary for the survival of the community, they also have access to power.

Paechter (2006c) also argues that 'power' has been significantly undertheorised within situated learning theory, something she attributes, at least partly, to Wenger's view of communities as uniformly benign. Paechter explores how 'power' functions in, and between, communities of practice. Her focus is on communities of gendered practice, but I will here limit the discussion to more general points raised by Paechter. Like Salminen-Karlsson (2006), Paechter's (2006c) preferred approach to power is largely Foucaldian, drawing on how Foucault's conceptualisation of power is intimately associated with knowledge. In particular, Paechter is interested in the possibilities to analyse how power (and resistance) operates on the micro-level that is opened up by the communities of practice perspective; how power/knowledge relations are expressed and resisted at particular times and places in particular communities. Thus, Paechter is advocating for an examination of power relations from within the communities of practice perspective, rather than, for example, the inclusion of a concept of hierarchy as advocated by Davies (2005), or effects of external power structures. In particular Paechter argues that issues of power need to be taken into account in relation to different individuals' possibilities to negotiate meaning within a community of practice:

> ...as different individuals within a group come from different positions, this negotiation can never be power-neutral, and one's value to the community affects what is actually negotiable. Those who are at the centre are far more likely to be able to ensure that their preferred meanings prevail. Furthermore, it is these people who have most to lose from change, and most to gain from preservation of the status quo.
>
> (Paechter 2006c, p. 20)

In this context it is important to take into account how power is never exclusively repressive, but is always both repressive and productive. This can be illustrated with an example borrowed from Ambjörnsson (2006), which I have transferred into the context of physics learning and expressed in the language of situated learning theory: Even when there is little room for negotiation of meaning within a community of practice this is not necessarily just repressive, it can also give students a feeling of control; they know what is expected of them and what it takes to be seen as a 'good' physics student. This understanding of power closely resembles Wenger's (1998) description of how power is derived from belonging to a community as well as exercising control over that community:

> On the one hand, it is the power to belong, to be a certain person, to claim the legitimacy of membership; and on the other it is the vulnerability of belonging to, identifying with, and being part of some communities that contribute to defining who we are and thus have a hold on us.
>
> (Wenger 1998, p. 207)

In summary, I would agree with Contu and Willmot (2003), and argue that situated learning's key notion of learning, which involves the constitution of identities (and not merely an acquisition of knowledge or skills) does in fact open up interesting possibilities for an inclusion of power relations as an important element in this theory of learning.

3.2 Post-Structural Gender Theory

From a situated learning perspective participation in a practice is the key to 'being' and 'becoming', for example, a physicist. You cannot simply claim to be a participant in a community of practice, in order to uphold the 'membership' a continuous participation in (and negotiation of) the practice is required. While it may be reasonably straightforward to accept that, in order to be recognised as a physicist you need to 'perform' physicist in 'correct' ways, it may be somewhat more provoking to think about 'gender' in such terms – but this is precisely what feminist post-structural theory challenges us to do. Paechter (2003a) explains:

> Identity can in this way be seen as being related to competent and convincing performance of a particular role; it is defined not just internally by the individual but externally by the group's inclusive or exclusive attitude to that individual. … Thus, it becomes not sufficient to claim a particular identity; that identity has to be recognised by group members, which in turn reflects back on one's understanding of oneself.
>
> (Paechter 2003a, p. 74)

At the core of feminist post-structural theory is a 'troubling' of 'the very categories of male and female to make visible the way they are constituted and to question their inevitability' (Davies and Gannon 2005, p. 318). In short, gender is portrayed as something fluid, something continuously changing, not an inherent characteristic of a person. Hey (2006) describes this as follows:

> The central poststructuralist ideas that the subject is an effect rather than a cause is the key to Butler's theories of performative identities. Deconstruction is thus a form of critique focused on examining the role of discourse in asserting forms of identity.
>
> (Hey 2006, p. 444)

Butler (1999) elaborates further:

> In this sense, gender is not a noun, but neither a set of free-floating attributes, for we have seen that the substantive effect of gender is performatively produced and compelled by the regulatory practice and gender coherence. Hence, within the inherited discourse of the metaphysics of substance, gender proves to be performative – that is, constituting the identity it is purported to be. In this sense, gender is always a doing, though not doing by a subject who might be said to pre-exist the deed… There is no gender identity behind the expressions of gender; that identity is performatively constituted by the very 'expressions' that are said to be its results.
>
> (Butler 1999, p. 33)

To illustrate what performing of gender can mean in practice I have borrowed an example from Ambjörnsson (2004). She has studied how gender (as well as sexuality and ethnicity) is done by girls in secondary school; this is one way they can perform a certain kind of femininity:

> To walk down the school corridor with your book pressed against your chest, sit down at the bench next to another girl and giggling lean your head towards hers, are thus actions that in themselves create gender. However, it is not enough to once and for all giggle with girlfriends, dress in a skirt and put up your hair in a pony-tail. Gender has to be recreated continuously in order to be convincing. And it is this recreation – this eternal repetition – that means that gender cannot be viewed as a static state. Rather it must be viewed as a verb, a continuous present tense – a process.
>
> (Ambjörnsson 2004, p. 12-13, my translation from the original Swedish).

In other words, it is only through a repeated successful performance that a person can attain identity in a given context as man or woman. Moreover, the view of gender as performative is, for Butler (1999), a way to break down gender binaries, to allow for a wider variety of possible ways of doing gender:

> The reconceptualization of identity as an *effect*, that is, as produced or generated, opens up possibilities of 'agency' that are insidiously foreclosed by positions that take identity categories as foundational and fixed.
>
> (Butler 1999, p. 187)

In going beyond the dualistic view of gender as masculinity versus femininity the notion of multiple masculinities and femininities becomes important (Connell 2003). Here Connell notes that even the existence of the terms masculine and feminine indicates that we do conceptualize gender in terms of a range of practices, otherwise we would not need these terms and would only need to talk about men versus women, or possibly male versus female. Furthermore, the realisation that gender can be performed in a variety of ways open up a way for us to move away from a rigid dualistic view of gender. On the same note, Davies and Gannon (2005) stress that the task of feminist post-structuralism is 'not to document differences between men and women, but to multiply possibilities, to demassify ways of thinking about "male" and "female"' (Davies and Gannon 2005, p. 319).

Further, it is important to emphasise how feminist post-structuralism seeks to transcend the individual/social divide (Davies and Gannon, 2005), but in doing so is concerned with 'how individuals, within *specific social settings*, create and negotiate gender' (Sowell 2004, p. 26, emphasis added). Paechter (2007b) elaborates on the issue:

> The performance of masculinities and femininities is also, of course, a performance to and for others. Here, even more than to the self, it is successful performance that matters; even transgressive performances only have any point if they are read as such. Consequently, the performance of gender is a reciprocal relation between performer and audience, and its meaning will be interpreted in the relationship between them.
>
> (Paechter 2007b, p. 16)

Thus, stating that gender is performative does not mean that all kinds of 'performances' are possible; subjects can only take up those positions that are available to them (Honan et al. 2006). Paechter (2007b) explains: 'masculinities and femininities, while performative in nature, are not arbitrary; what can be performed is highly dependent on time, place and circumstances, including the power/knowledge relations in a specific context' (p. 40). But, as pointed out by Davies (2006) it is important not to confuse the importance of external powers with a passive, deterministic shaping of the subjects. Subjectification, as conceptualised by Butler, involves a simultaneous imposition and active taking-up of available positions (Davies 2006). For example, in their poststructuralist analysis of the schoolgirl, Hannah, Honan et al. (2006) are able to demonstrate how Hannah, as subjected to discourses of 'model female student and daughter', is able to actively take up and transform those positions to her own advantage. Furthermore:

> ...poststructuralist theory shows how it is that power works not just to force us into particular ways of being but to make those ways of being desirable such that we actively take them up as our own.
>
> (Davies and Gannon 2005, p. 318)

Thus, the subject is not only inscribed with values and norms from the outside, but those values and norms are also made desirable to adopt as our own, since they make us into recognizable and legitimate members of a social group (Davies and Gannon 2005).

3.2.1 Masculinities and Femininities as Communities of Practice

Assuming that gender can be understood as performative: How do we come to perform particular genders at particular times? This question is the starting point of an explorative article where Paechter (2003a) investigates how the idea of communities of practice can be used to think about the formation and perpetuation of masculinities and femininities:

> I am arguing that the learning of what it means to be man or woman within a social configuration results in shared practices in pursuit of the common goal of sustaining particular localised masculine and feminine identities. It follows from this notion that the localised masculinities and femininities within which these identities are developed and sustained can be seen as communities of practice.
>
> (Paechter 2003a, p. 71)

Paechter's article revolves around the question: 'What happens if we treat masculinities and femininities as communities of practice?' and it is in principal an examination of Wenger's characterisation of communities of practice where she demonstrates how masculinities and femininities can be treated this way.

At the core of her argument is the shared focus on practice/doing within communities of practice and post-structural gender theory; how the enactment of specific practices is what grants a person membership in a community of practice as well as what allows one to become recognised in terms of performing a particular masculinity or femininity. In short, how identity is 'understood through the practice in which we engage' (Paechter 2003a, p. 74). As with any other kind of community of practice, communities of masculine and feminine practices are involved in the constant production, reproduction, and negotiation of what it means to be, in this case, a man or a woman. One important advantage of treating masculinities and femininities as communities of practice is, Paechter (2003a) argues, that it allows one to take into account individuals' many overlapping and context-dependent masculinities and femininities and thus provides the tools to analyse the interrelationships between these different participations. Further, Paechter (2006) also sees a possibility that a conceptualisation of masculinities and femininities as communities of practice can contribute to an undermining of some of the binaries that are associated with traditional approaches to gender, since such a conceptualisation:

> ...allows for fluid boundaries across and between different masculinities and femininities; an individual does not have to be committed to one but can take up/perform/inhabit several, at different places and times.
>
> (Paechter 2006, p. 14)

What masculinities and femininities we perform will thus be connected to the situation, but moreover, they will also be understood and read differently depending on the situation – what is considered masculine in one community may just as well be considered feminine in another, even by the same person. This reflects not only how localised masculinities and femininities can be, but also how they are defined in relation to one another; when something is conceptualised as 'masculine' this is always done in relation to something else that is conceptualised as 'feminine', these conceptions do not exist independently of each other (Paechter 2007b).

Primarily Paechter (2003a, 2003b, 2007b) is interested in how young children gradually learn what it is to be male and female within particular contexts, how they can be viewed as legitimate peripheral participants in adult communities of masculinities and femininities, learning their shared repertoire of performance. Paechter (2003a) exemplifies:

> Boys engaging in legitimate peripheral participation in some adult communities of masculine practice may, for example, learn that one does not cry, and that personal hurt is not talked about, while the public expression of anger is perfectly OK.
>
> (Paechter 2003, p. 72)

However, I believe that viewing masculinities and femininities as communities of practice is also useful for understanding how the doing of gender – thus, participation in particular masculinities and femininities – is intertwined with the participation in professional communities of practice, perhaps in particular ones that have strong gendered connotations, such as the physicist community of practice.

Finally, it needs to be acknowledged that using to terms 'masculinities' and 'femininities' is far from unproblematic, something Paechter (2006b, 2007b) also discusses. By referring to certain practice as a masculinity and (others) as a femininity there is a danger in reproducing masculine-feminine dichotomies. Further, there can be tendencies to, as pointed out by Gilbert and Calvert (2003), assume a 'close relationship between actual men and "masculinity" and between actual women and "femininity"' (p. 875); to interpret women's practices as femininities and men's as masculinities. To at least some extent, the first of these considerations can be met by taking multiple masculinities and femininities into account. The second concern is more difficult to resolve, but Paechter (2007a) has suggested that:

> It may be that, despite what would be bound to be a problematic terminology, that we should think of masculinities in terms of power/knowledge that are traditionally, but not always successfully, claimed by men, and make the case that these are also mobilised by women and girls, rather than focusing on what men actually do in particular local circumstances.
>
> (Paechter 2007a, p. 10)

Following this suggestion I decided to, for example, talk about a 'traditional science student femininity', a femininity that incorporates the characteristics previous research often has associated with woman science students (see section 8.6.1.2).

3.3 Concluding Remarks

In this chapter the theoretical basis of my research, that of situated learning theory and post-structural gender theory, has been introduced. In doing so I have discussed how themes central to my research, such as learning and gender, are conceptualised within these theories. I have also indicated how the theories share common ground in their respective conceptualisations of physics and gender as situated doings.

In the next chapter the broader theoretical staging of this chapter is sharpened into my theoretical framework; a development of Wenger's (1998) communities of practice perspective that incorporates a theory about gender.

CHAPTER 4

Theoretical Framework

'There is nothing as practical as a good theory'
Lewin (1952)

4.1 Introduction

This chapter[12] takes as its vantage point the theoretical purpose of the thesis; *to formulate a framework that allows for the exploration of physics students' gendered identity constitution in, and beyond, my empirical study*. As pointed out in the last chapter this theoretical framework was developed in close association with my analysis of the transcripts from the interviews with the undergraduate students. Thus, in short, the theoretical framework is a product of the meeting between situated learning theory, post-structural gender theory (both introduced in the previous chapter) and my empirical material.

In a sense, the broader theoretical basis of my research presented in the previous chapter influences the entire research process, from the formulation of the research purpose to the final analysis. But, when working with the empirical material in the actual analysis, more focused 'tools of inquiry' are needed. In the case of my research these tools of inquiry are of two kinds; analytical and theoretical. In this chapter the theoretical framework is introduced. Essentially the theoretical framework is woven around Wenger's (1998) conceptualisation of a community of practice, but in order to allow for an analysis of the gendering of the students' identity constitution several aspects have been added. Firstly, I draw on Paechter's (2003a) conceptualisation of masculinities and femininities as communities of practice – in short, gender as a community of practice. Secondly, insights from, for example, sociological and anthropological studies regarding the gendering of physics are included – gender in the physicist community of practice. Thirdly, I let the communities of practice perspective meet my empirical

[12] A version of this chapter has been published in Gender and Education (Danielsson and Linder 2009).

material in such a way that it brings out the gendered aspects of the doing of physics – how the students simultaneously do gender and do physics. As such the theoretical framework can be understood as a sharpening of the theoretical staging presented in the previous chapter; while the theoretical staging is concerned with a broad characterisation of concepts fundamental to my research such as 'gender' and 'learning', the aim of the theoretical framework is to provide a more focused 'tool of inquiry' for the analysis of students' simultaneous constitution of (a gendered) practice and (a gendered) identity.

4.2 Structure of the Chapter

In what follows, I elaborate on how Wenger's notions of practice and identity inform my theoretical framework. In doing so, excerpts from the interview with the undergraduate student 'Ann' are presented alongside the aspects of the theoretical framework used to interpret the excerpt, thereby simultaneously illuminating Ann's experiences and the application of my theoretical framework.[13] The analytical outcomes for the research purpose of the thesis, which draws on all of Part II of the thesis – the full complexity of the conceptual framework – is presented in Part III.

4.3 The Practice of Physics

A shared practice is, Wenger (1998) argues, what holds a community together. This practice can be understood as being composed of a number of dimensions (meaning, community, learning, boundary and locality). By exploring how these different dimensions can be understood as being influenced by gender – some more and some less – it is possible to form a comprehensive picture of the gendering of the practice of physics. Thus, this perspective serves as a useful lens for analysing gender in the community of practice of physicists, an understanding that then will be helpful for the exploration of how students constitute their identities in relation to that community.

Lave and Wenger (1991) (and later also Wenger 1998) treat communities of practice as highly localised entities, an approach that has been criticized by, for example, Nespor (1994). Nespor counter argues that communities need to be treated, not as bounded, strictly local settings, as he claims Lave and Wenger (1991) are doing, but as parts of an 'actor-network' that links activities across space and time. In particular, Nespor (1994) argues, it 're-

[13] There are a few occasions where excerpts from other interviews are also used to illustrate various aspects of the theoretical framework, as all aspects of the framework of course are not applicable to a single interview.

quires us to look closely at how distant activity is transported into and made manifest in the particular settings, and how activities in those settings are connected to activities and spaces elsewhere' (p. 3). In principal I agree with Nespor's critique, but still think that the concept of communities of practice is a useful one to come to better understand aspects of the learning of physics, albeit when used in a somewhat less strict sense. In my understanding, Nespor's actor-network theory and situated learning theory serve somewhat different purposes. Nespor's actor-network theory focuses heavily on creating a structural understanding of a 'community', lacking a deeper understanding of the individual and their agency. While this is by no means included per se in a communities-of-practice-perspective, I do see possibilities for such an application of the perspective. In particular, when complemented with theoretical insights from Nespor's work, such as 'the importance of including learners' of a discipline, such as physics, into the actual community of physics, since even though students 'aren't doing "real" physics [this] doesn't mean that their doing isn't an essential part of the trajectory to doing 'real' physics' (Nespor 1994, p. 48). Furthermore, in order to more fully understand an individual's joining of a community of practice a more profound way of understanding the workings of power relations in and between communities needs to be developed, as I have discussed in section 3.1.2.

As demonstrated by Davies (2005) defining a community of practice, and in particular defining who is a member, is far from straightforward. I have, however, found it most useful for my purposes to start off with a relatively loose definition of the term, since the students' education opens up many possible trajectories into (and out of) the world of physics. The students in this study should not be seen as legitimate peripheral participants in terms of only one particular, localized physicist community, but should rather be seen as participants in the broader sense of the physicist community as a whole (I discuss this further in section 8.5). I do acknowledge that the broader physicist community of practice is a highly complex one, consisting of many interrelated communities, one of which is the laboratory-based physics-student community. However, in line with the research purpose, it is not fruitful to provide a further distinction between the different possible physicist communities. I would argue that for now it is sufficient to note that the communities that make up physics, as an essentially empirically driven science, all have some strong links to laboratory settings. These settings, in turn, are widely seen to be an essential and integral part of learning to become a physicist. The definition of the physicist community of practice will be further explored in section 8.5, where I, starting from the interviews with the graduate students, ask the question whether the university-based physicist community of practice is best understood as one or several communities. Thus, at this stage, it is sufficient to work with a less rigorous definition of the physicist community of practice and leave the more fine-grained exploration of the structure and boundaries of the community to the empirical investigation.

4.3.1 Meaning

Central to Wenger's (1998) concept of practice is 'meaning'. Meaning, as understood by Wenger, is located in the negotiation of meaning. He defines this as 'the process by which we experience the world and our engagement in it as meaningful.' (p. 53). He further elaborates:

> The negotiation of meaning is a productive process, but negotiating meaning is not constructing it from scratch. Meaning is not pre-existing, but neither is it simply made up. Negotiating meaning is at once both historical and dynamic, contextual and unique.
>
> (Wenger 1998, p. 54)

In other words, the concept of meaning negotiation captures the dynamic relation of living in the world; people do not make meaning totally independent of the world, but meaning at the same time is not just imposed on them by the world. The negotiation of meaning involves the two interacting processes 'participation' and 'reification'. Participation means the members actively taking part in the community and their relation to others is reflected in this process. Wenger (1998) writes 'I will use the term participation to describe the social experience of living in the world in terms of membership in social communities and active involvement in social enterprises' (p. 55). Reification, on the other hand, 'refers to the process of giving form to our experiences by producing objects that congeal this experience into "thingness"' (Wenger 1998, p. 58), thus creating something 'real' to organize our negotiation of meaning around. Wenger goes on to describe the process of reification as central to every practice, producing abstractions, tools, symbols, stories, terms, and concepts. Moreover, participation is something that goes beyond our mere engagement in practice. You do not cease to be a physicist, for example, when you leave your laboratory to go home. Looking at how gender is intertwined into the participation and reification of the physicist community opens the way for me to explore how the negotiation of meaning can be understood as gendered.

To clarify the dynamics of negotiated meanings as well as the concepts of participation and reification I have transferred an example given by Wenger (1998, p. 54-55) into the context of doing laboratory work in physics: A student doing a laboratory exercise in physics can be understood as an example of negotiation of meaning. The doing of the laboratory exercise is situated in a complex context, that includes, for example, the student's educational background, the laboratory instructions, the student's experiences with similar exercises, who the other students and teaching assistants are, etcetera. All these things contribute to shaping the student's experience of this particular exercise. When a physics student steps into the student laboratory this situation is both well-known and new. She brings with her previous experiences of doing laboratory work; common problems and annoyances as well as ways of handling them. The laboratory exercise as such also carries a history. It has been shaped and reshaped by teachers and teaching assistants in response to changing pedagogical, technological and scientific developments and demands. The physics student contributes to the negotiation of meaning by being a member of the community and by bringing with her her history of participating in the practice. In a similar fashion, the laboratory exercise contributes to the negotiation of meaning by its reflection of aspects of the practice. The physics student as a member of the community of practice here embodies a long and diverse process of what Wenger calls 'participation'. Similarly, the laboratory exercise 'as an artefact of certain practices embodies a long a diverse process of [reification]'. Further, 'it is in the convergence of these two processes in the act of [doing the laboratory exercise] that the negotiation of meaning takes place'.

How a student's participation can be gendered is described by Calabrese Barton (1997). As an undergraduate she did not feel confident nor comfortable with quantum mechanics, seeing it as a subject reserved for 'geniuses' and attributing her good grades to 'pure luck'. However, she claims that her study of feminist theories of science made her understand that her discomfort with the world-view of physics was not rooted in a lack of intelligence, but rather a consequence of her attempts to 'engage in a world that historically [had] not appreciated or even respected the beliefs and values [she] had learned to value as a woman' (Calabrese Barton 1997, p. 148).

4.3.2 Communities and Boundaries

The community formation in terms of practice is depicted by Wenger (1998) as practitioners' mutual engagement in a joint enterprise resulting in a 'shared repertoire'. The negotiation of the joint enterprise gives rise to mutual agreement on, for example, what matters for the community and what does not, what is important and what is not important, what to pay attention to and what to ignore. These in turn result in a shared repertoire of the community that includes things such as words, tools, ways of doing things, stories, and symbols. This repertoire of the community is shared in a dynamic and interactive sense among the members as a resource for the negotiation of meaning. How the shared repertoire of science can be seen as gendered has been explored by, for example, Merchant (1984) and Calabrese Barton (1997). Calabrese Barton (1997) has criticised the language of science for being unemotional, competitive and aggressive – characteristics all associated with hegemonic masculinity. Merchant (1984) points out how the symbolic structure in science is permeated with ideas about gender and how these symbols, these metaphors, do in fact have a normative function for the formulation of what science is. In particular, Merchant emphasizes the identification of nature with a woman, often a woman carrying secrets, as one of the strongest of these symbolisms: a symbolism that throughout history has been used by scientists when they discuss their research.

When considering students' joining of a community I would argue that, in particular, an understanding of the shared repertoire of physics, or rather science in general, is crucial. It should be borne in mind that these students as legitimate *peripheral* participants in the physicist community have very limited possibilities of actually influencing the shared repertoire (see also sections 4.3 and 8.5). In a broader sense, the most important component of the shared repertoire of science is the delineation of what is seen as belonging to science; what is considered scientific and how this has developed historically – most certainly a gendered process, as demonstrated by, for example, Schiebinger (1991).

During the interview with Ann, when asked how she works in the student laboratory, she expressed the following:

> Ann: 'Cause most of the time I've been doing lab work with guys and then most often I've taken the female role, partly because I feel a bit slow.[14]

To understand this description I find the notion of 'boundaries' (between communities) illuminating. Whereas the dimension 'community' describes what belongs to physics, it is equally important for its definition to look at what does not belong – how the boundaries of the physics community are defined and in particular how the community is related to other communities of practice. No community of practice ever exists in isolation and consequently cannot be understood independently of other communities. This implies that when one joins a community of practice one engages not only in the practice as such, but also in this community's relations with the rest of the world. Furthermore, communities of practice can be seen as shared histories of learning, and over time these histories create discontinuities between those who have participated and those who have not. It is through one such discontinuity that Ann's above description can be interpreted; that between femininity and science. Drawing on a discourse of femininity and science being defined in opposition to one another (see, for example, Schiebinger 1991 and Brickhouse 2001), 'female role' is here expressed as being related to 'being slow' and thus being incompatible with participation in science – a discontinuity, a boundary is constituted between science and femininity. I will return to this perceived discontinuity in the next section as I further explore how Ann positions herself as a particular kind of woman and a particular kind of physicist.

Such discontinuities can stem both from participation and reification. In some cases, for example, a boundary of a community of practice can be rei-

[14] The interview was conducted in Swedish; the English translation captures the meaning of the Swedish original rather than being a literal translation.

fied with explicit markers of membership, such as titles. While this might seem to be the case for the physicist community of practice – the boundary being reified by having a degree in physics – it is demonstrated in sections 8.4 and 8.5 that the definition of the boundary of the physicist community of practice is a lot more complex than that. It also needs to be noted that the absence of explicit markers in no way implies the absence or looseness of a boundary. However, participation and reification can also create continuities across boundaries. Wenger uses the term 'boundary object' to describe how a product of reification can be used in several communities and thus create a continuity between the communities. For the undergraduate student Kalle in the interview excerpt below, the workshop (at least in a metaphoric sense) functions as such a boundary object. Prior to his physics studies, Kalle worked in a workshop and in explaining his participation in the physics student laboratory he says:

Kalle: But it is precisely that that's so appealing, that it's so close to working in a workshop really…

Thus, Kalle is able to create a sense of continuity between his previous participation in a workshop community and his present participation in a physicist community by drawing on the existence of 'workshops' in both. Not only artefacts, but also people can create continuities between communities by their participation in multiple communities. Wenger calls this use of multimembership to transfer elements of one practice into another 'brokering'.

4.4 Identity in Practice – Doing Masculinities and Femininities in Physics

Issues of identity are, according to Wenger (1998), 'an integral aspect of a social theory of learning and are thus inseparable from issues of practice, community, and meaning' (p. 145). By changing the perspective on the community of practice from 'practice' to 'identity' it is possible to simultaneously zoom in on the individual and extend the view out beyond the community, including the individual's engagement in a larger societal context. Thus, this perspective allows me to consider individual students' engagement with the gendered practice of physics, analyzing how they constitute their identities as physicists in relation to its gendered practice, while doing laboratory work.

Whereas the different dimensions of practice are helpful in understanding how the practice of physics is gendered, the concept of 'identity in practice' is what relates the individual student's identity constitution to this practice. Consequently, identity is first of all seen as a negotiated experience, not a stable category. In other words, our identity is something we continuously constitute. Thus, identity is a constant 'becoming', something that is always going on. Therefore, our identities can be viewed as forming trajectories, both within and across communities of practice, and this formation is what learning is all about:

> Understanding something new is not just a local act of learning. Rather, each is an event on a trajectory through which they give meaning to their engagement in practice in terms of the identity they are developing.
>
> (Wenger 1998, p. 155)

This understanding of identity has much in common with a post-structural understanding of gender, as discussed in section 3.1.1, in that both situated learning theory and post-structural gender theory put such a strong focus on practice; it is our doings that create our identity/our gender. As Brickhouse (2001) also points out:

> The idea of learning as a *transformation of identity-in-practice* provides a way of thinking about learning that is gendered, but does not regard gender as a stable, uniform, single attribute. We are not born with gender. We do gender.
>
> (Brickhouse 2001, p. 290, emphasis added)

Taken together, situated learning theory's notion of learning as identity formation; the way in which we craft our identity through engagement in a practice; and, the notion of gender as a community of practice, provide an entirely new way of understanding and analysing students' participation in a gendered discipline, such as physics. This is because, as the students participate in practices in the student laboratory, not only do they craft their identities as legitimate peripheral participants in the physicist community, they also craft certain masculinities and femininities, in relation to the practice. And, the students are also crafted by the practice; as participants in a community of practice they are subjected to the norms and expectations of the community.

In the case of Ann, I would like to return to her previous mentioning about how she takes 'the female role' because she experiences herself as unskilled at laboratory work. This can be understood as her way of doing femininity in relation to the demands of the practice of physics. Further, consider the following excerpt from the interview with Ann:

> Ann: If you are to do labs, put some stuff in and then measure the result, that fits me really well and then analyse it, but not if you have to tinker too much yourself, try new things, and connect stuff together and such, that doesn't fit me.

This, I would argue, can be understood as Ann doing gender by distancing herself from what could be characterised as a 'practical masculinity' (Wajcman 1991, Mellström 1999). Wajcman (1991) has characterised the gendering of technology in terms of two masculinities, one practical based on physical strength and machine-related skills and one based on 'the professionalized, calculative rationality of the technical specialist' (p. 144). These dual masculinities of technology have since been applied by Mellström (1999) in his research about engineers. I also find them useful for understanding students' descriptions about 'doing physicist', albeit with the tradeoff that there is a risk of 'essentialising', that is of portraying certain characteristics as being inherently masculine. By distancing herself from the 'practical' masculinity Ann distances herself from the kind of masculinity that characterised her previous workplace, the electrical workshop, without challenging the gender norms of physics. This distancing takes place both by dissociation from certain types of laboratory work (as in the previous interview excerpt) and by Ann positioning herself as 'different' from coursemates representing more of a practical masculinity, such as Mats:

> Ann: Mats, he's an experimentalist! He's so much fun to do labwork with, 'cause he really gets, he might not understand the theory at all and hasn't done anything and is tired and hasn't slept and he sure starts to tinker kind of! He's so very different [from me], he really fits in a lab!

Furthermore, work of identity transformation will of course also be influenced by the students' simultaneous belonging to other (sometimes gendered) communities of practice. These multiple participations have been discussed by Wenger (1998) in terms of 'a nexus of multimembership'. Following Wenger an identity can be viewed as a nexus of multimembership, a concept that brings our multiple belonging to many communities of practice to the fore: 'Our various forms of participation delineate pieces of a puzzle we put together rather than sharp boundaries between disconnected parts of ourselves' (Wenger 1998, p. 159). In other words, the belonging to a community of practice is not something one can turn on and off, and thus, this belonging will influence our belonging to other communities of practice. For example, our belonging to particular, perhaps class-based, masculinities and femininities will affect the way we can join the physicist community – we cannot just turn off those other belongings when we step into the physics student laboratory. Furthermore, in order to maintain one's identity across boundaries, reconciliation is necessary. As further elaborated by Wenger (1998): 'Different practices can make competing demands that are difficult to combine to an experience that corresponds to a single identity' (Wenger 1998, p. 159). How students deal with such competing demands is an important theme in the analysis, in particular regarding individual students' struggles to negotiate their participation in certain gender manifestations with their participation in the physics student community.

I interpret Ann's description as partly constituting her identity as a physicist through participation in what could be characterised as a 'traditional science student femininity' (this is further elaborated on in section 8.6.1.2). Firstly, throughout the interview she comes back to how her main focus in the student laboratory is the discussions surrounding the laboratory work rather than the actual 'doing', thereby reproducing a rather typical femininity, as, for example, in the following excerpt:

I: Has there been any difference between, if you think about the introductory mechanics course compared to the nuclear and particle physics now?

Ann: Yes, they've been great. Now [inaudible] has wished that they've had a lesson before and had a proper introduction. Often when you get to a lab and then you're not really in phase with the course and you haven't had time to read what you're supposed to and then you stand there like a fool when you're supposed to do things, but here they really had half an hours introduction and went through all the theory we had in the lab and that was great. And then... on one hand you understand more what you're doing, on the other hand all the concepts come together.

I: That you understand the theory 'cause now you see... 'Cause now you talk about it, not because you see it, or..?

Ann: No, because you talk about it all the time, at least that's what I think.

Secondly, the above excerpt voices a need to be well-prepared, another aspect that is often attributed to woman science students. Here previous research has described woman students as, in the context of learning science, 'doing what they are told', whereas the man students have a more playful approach to science and technology (Hasse 2002a; Jones et al. 2000; Mellström 1999). Ann also told of how a lecturer, in relation to doing laboratory work in electricity and magnetism, had said that the laboratory exercises were often easier for the man students, but that the women usually compensated for their possible shortcomings by being more thorough. In physics in particular, women's rule-following has been understood as a way of compensating for lack of earlier scientific and technological experiences (Nair and Majetich 1995). However, I believe that there is more to the story than a lack of previous experiences; with the underlying ideology of science having been interpreted as masculine (Kleinman 1998), I agree with Byrne (1993) in her claim that the masculine, disciplinary culture of science can be an important barrier for those adolescent girls, who want to be seen as 'feminine', to pursue a career in science. Consequently, the rule-following among some women participating in science could be interpreted as a reconciliation between 'doing physics' and 'doing femininity', by drawing on a discourse of women being more thorough:

Ann: I think I benefit from being quite structured, I can take… It's almost the same thing; that first I do this and then I do that. And that you've learnt along the way that you bring that to the labroom too, that you… that you do one thing at the time and that you kind of… you reconnoitre first, ok, what am I supposed to do, what apparatus should I use and how do I do that… you're really careful throughout all the steps.

Related to this thoroughness, this rule-following, is also the work after the actual laboratory exercise, the writing of the report. When Ann is asked what she sees herself as good at when it comes to doing laboratory work she says:

Ann: I can be structured beforehand, that I can… I've tried to go through stuff before and then I think I'm pretty good at writing reports, that I'm good at the work afterwards. When it comes to pure labwork it's those things I'm good at.

I would argue that this focus on writing abilities can be understood as another aspect of Ann's doing of 'traditional science student femininity'. The writing in the student laboratory is most often associated with the woman students (Rosser 1995), and by claiming proficiency in this area Ann is able to participate in physics by a doing of femininity.

4.5 Participation and Non-Participation

I have so far looked at how identity and practice are interrelated; but it is not only through participation in certain practices that identities are formed. Equally important in fact, can be the non-participation in other communities. By constituting identities not only by what one is, but also by what one is not, non-participation becomes just as much a source of identity as participation. This becomes especially crucial when looking at communities who define themselves by contrasting themselves to others; where being inside one community is the same as being outside another – for example, workers versus managers or different ethnic groups. In particular, this can make 'boundary crossing difficult because each side is defined in opposition to the other and membership in one community implies marginalization in another' (Wenger 1998, p 168). By exploring how students talk about their (and others) participation in the physics student community versus how they describe their participation in various masculinities and femininities it is possible to create an understanding of how these participations and non-participations are interconnected. What masculinities and femininities are, for example, understood in opposition to, or alignment with the physics student community?

In Ann's identity constitution the relationship between femininity and physics is central. For example, she says:

> Ann: I can never be like normal [women] [...] So I feel very comfortable among guys...

To understand how Ann positions herself within the relationship between femininity and physics the notions of participation and non-participation are useful. Throughout the interview she comes back to how she is not a traditional woman and how she feels more comfortable in man-dominated than all-woman environments. This positioning as a non-participation in a traditional femininity is common for many women within science and technology; they tend to explain their presence in such a masculine subject by constructing themselves as different from other women, i.e., as 'being one of the boys' – as participating in a masculinity (see, for example, Hughes 2001, Henwood 1998 and Walker 2001). I would argue that this perceived non-participation in a traditional femininity needs to be understood in relation to these women's participation in physics, or science and technology in general. It can be interpreted as a result of the historical discontinuity between femininity and science as, for example, described by Schiebinger (1991). Schiebinger argues that what has been seen as feminine and what has been seen as proper science has historically been defined in opposition; what was scientific was by default unfeminine and characteristics seen as feminine were not needed nor wanted within science. Consequently the categories 'feminine' and 'proper science' are almost mutually exclusive. In other words, participating in both femininity and science was just not possible, an idea that is still prevalent among today's woman physics students.

In summary, the argument here is that 'traditional femininity' and 'physics' need to be understood together in order to comprehend how, in particular, woman physics students constitute their physicist identities. Ann is thus drawing on the discontinuity between 'traditional femininity' and physics in order to position herself as non-participating in one and participating in the other.

Is, however, this non-participation in a traditional femininity a contradiction to Ann's earlier claim of taking the 'female role' in the student laboratory? In a sense it is, but this should come as no surprise as our choice of subject positions are highly context-dependent and vary between situations. We therefore need to take into account multiple and sometimes conflicting gender manifestations, since, as pointed out by Paechter (2006c) 'different identities may ... be claimed and performed in the different locations' (p. 22). This is nicely illustrated by the following excerpt from the interview with Ann, where she, despite repeatedly having positioned herself as unskilled in the practical work in the student laboratory, says the following:

> Ann: I think I'm untraditional. Because on one hand I've worked in that workshop, well, I wasn't so very, I wasn't the one who got the more difficult jobs, but I worked there quite a lot, you know you stand there and connect stuff. ... So

> I'm pretty handy when it comes to such stuff. Have changed the brake shoes on the car once, just so that I should have done it sort of. Now I've done it, will never do it again.

Thus, it is necessary to consider an individual's participation in various communities as something relatively fluid, not the fixed 'membership' Wenger has been criticized for applying (see, for example, Nespor 1994). So, while communities of practice might be quite resistant to change, individuals participation in them are not necessarily so, opening up possibilities for agency and change.

4.6 Modes of Belonging

In order to understand identity formation in relation to learning Wenger (1998) argues that engagement is not the only mode of belonging that needs consideration. This is because a person's identity formation is affected by social processes and configurations that reach beyond the person's direct engagement in practice. He describes three distinct modes of belonging:

> 1) engagement – active involvement in mutual processes of negotiation of meaning.
> 2) imagination – creating images of the world and seeing connections through time and space by extrapolating from our own experiences.
> 3) alignment – coordinating our energy and activities in order to fit within broader structures and contribute to broader enterprises.
>
> (Wenger 1998, p. 173-174)

On one hand, the engagement in practice gives a participant opportunities to negotiate the enterprise and transform the community as well as their identity. However, engagement can also be narrow:

> The understanding inherent in shared practice is not necessarily one that gives members access to the histories or relations with other practices that shape their own practice. Through engagement, competence can become so transparent, locally ingrained, and socially efficacious that it becomes insular: nothing else, no other viewpoint, can even register, let alone create a disturbance or a discontinuity that would spur the history of practice onward. In this way, a community of practice can become an obstacle to learning by entrapping us in its very power to sustain our identity.
>
> (Wenger 1998, p. 175)

When considering students joining a community of practice I find the concept of imagination particularly useful: it can play a powerful role when creating a sense of belonging in the physicist community. By extrapolating from our own experiences we can imagine the working lives of other people.

For example, how does a student experience the student laboratory in terms of resembling what professional physicists do? The work of imagination thus involves defining a trajectory that connects what we are doing, for example in the student laboratory, to an extended identity, for example, being a physicist. How a student does this can consequently make a big difference to the potential of learning through a particular activity: different students can learn very different things from the same activity. This is, then, in turn also interconnected with our other, sometime gendered belongings; to what extent can these hinder or facilitate our possibilities to imagine ourselves as participants in a particular community. In the case of the interviewed undergraduate student David his strong distancing from what could be characterised as a 'practical technological masculinity' makes it difficult for him to imagine himself as participating in the practice of a research laboratory. This example can also be used to illustrate one of the trade-offs of imagination, that it can be based on stereotypical assumptions about a practice. David's possibly stereotypical assumptions about the practice of the physics research laboratory thus close one possible trajectory for his continued engagement in the community.

4.7 Identification and Negotiability

In order to gain an understanding of identity constitution that is as complete as possible, Wenger (1998) stresses that it is needed not only to take 'identification' but also 'negotiability' into account. He writes:

> Our identities form in this kind of tension between our investment in various forms of belonging and our ability to negotiate the meanings that matter in that context.
>
> (Wenger 1998, p. 188)

Identification refers to the process of constituting our identity through the previously introduced modes of belonging. This is something we both do to ourselves and to each other. To become members of a community it is not sufficient to identify with the community, one also needs to be recognized by the other members of the community. Whom the students interviewed recognise as members of the physicist community is discussed in section 8.4. However, 'engagement may not be perceived as an obvious source of identity by those involved' (Wenger 1998), as:

> In the context of engagement, identification takes place in the doing; hence, the development of competence, the negotiation of a trajectory, and the work of reconciliation across boundaries are not necessarily self-conscious. Identification through engagement does not require a reification of the relations it is based on.
>
> (Wenger 1998, p. 193)

Negotiability concerns our possibilities to control the meanings within a community of practice in which we are invested. It refers to 'the ability, facility, and legitimacy to contribute to, take responsibility for, and shape the meanings that matter within a social configuration' (Wenger 1998, p. 197). How identification and negotiability are closely interconnected in a student's identity constitution is well illustrated by the excerpt below from the interview with the Master's student Hanna. She does not identify as a physicist and accordingly she does not experience herself to be able to negotiate what it means to be a physicist:

> I: Do you see yourself as a physicist then?
>
> Hanna: No, not really that either [laughter]. Actually. I've never thought about it like that.
>
> ...
>
> I: What is important to be skilled at then, as a physicist or an engineer or whatever you're on your way towards?
>
> Hanna: ... Well, it's to be able to independent or ... critically examine ... I don't know. Erm...

In summary, viewing identity constitution as a dual process of identification and negotiability gives rise to questions such as: What is needed in order to be identified as a physicist? To what extent do physics students perceive that it is possible for them to negotiate meanings within the physicist community of practice?

4.8 Concluding remarks

In this chapter I have presented the theoretical framework I developed through the meeting between my empirical material, situated learning theory and post-structural gender theory. The theoretical framework draws on Wenger's conceptualisation of a community of practice, but develops it in ways that allow for an explicit analysis of students' simultaneous constitution of a gendered practice and a gendered identity.

In the next chapter I introduce the analytical tools I use as a complement to the theoretical framework, as a bridge between this framework and the empirical material.

CHAPTER 5

Analytical Tools

5.1 Introduction

This thesis sets out to explore how physics students' simultaneously constitute the practice of physics as enacted in student and research laboratories and their physicist identities in relation to this practice. In particular the thesis focuses on how these constitutions can be understood as gendered. As discussed in Chapter 2, related studies have often focused extensively on either exploring the doing of physics or the doing of gender, an asymmetry that I show can be overcome through the development of a theoretical framework that merges situated learning theory and post-structural gender theory, presented in the previous chapter. In both these perspectives the 'doing' is central to the understanding of identity; 'physicist' as well as 'masculinities' and 'femininities' are something 'done' through the participation in certain practices. However, though these perspectives are theoretically appropriate for my purposes, I found them difficult to apply directly to the empirical material. In particular I found that Wenger's (1998) identity perspective did not allow for the kind of fine-grained analysis needed in order to capture the dynamic of the interview situation. I have therefore added two complementary analytical tools to the theoretical framework. These tools are drawn from positioning theory and Gee's Discourse analysis respectively.

As argued by Linehan and McCarthy (2000) positioning theory can provide an analytical complement to situated learning theory in that it allows for a more fine-grained analysis of the individual's relation to a community of practice. Gee's Discourse analysis is developed from numerous sources of inspiration, of which situated learning theory is an important one. In fact Gee's characterisation of a Discourse is in many ways similar to the characterisation of a community of practice. Consequently, the tools of inquiry developed by Gee are highly applicable for doing research from a situated learning perspective, and I have found one particular 'tool', that of 'Discourse models' useful for my purposes. In this chapter I describe the analytical tools of 'Discourse models' and 'positioning' and show how they relate to the theoretical staging of my research.

The analytical tools of Discourse models and positioning provide an entrance to the empirical material, allowing me to constitute narratives of aspects of the interviewed students' identity constitution (see Chapter 7). This applications of the analytical tools has allowed me to bring to the fore, for example, how the undergraduate student David in the interview is positioning the woman physics students and how the undergraduate student Mia in the interview associates herself with a particular Discourse model, claiming that this is the only one available to her as a woman. These narratives are then theoretically deepened by the application of the theoretical framework presented in the previous chapter.

5.2 Discourse Models

Discourse analysis is a common tool within qualitative educational research (see, for example Northedge 2002 and Wickman and Östman 2002). What characterises Gee's (2005) 'big D' Discourse analysis is that it takes a holistic approach to the discourse. Gee writes, 'when "little d" discourse (language-in-use) is melded integrally with non-language "stuff" to enact specific identities and activities, then I say that "big D" Discourses are involved' (Gee 2005, p. 7). Gee's Discourse can thus be characterised as including the broader societal context in which the language is used; where, for example, gender becomes an important component. In summary, Gee (2005) views us as building and rebuilding our world through language used together with 'actions, interactions, non-linguistic symbol systems, objects, tools, technologies, and distinctive ways of thinking, valuing, feeling, and believing' (Gee 2005, p. 10).

Since identity is central to Gee's Discourse analysis, from this perspective learning is understood as involving a process of identity formation (Case and Marshall 2006). I would further argue that with Gee's wider definition of discourse, entering of a discourse can be seen to be linked to participating in a community of practice. This becomes apparent, for example, as Gee (2005) explains how Discourse allows one to be recognized as a particular kind of person:

> The key to Discourse is 'recognition', If you put language, action, interaction, values, beliefs, symbols, objects, tools, and places together in such a way that others *recognize* you as a particular type of who (identity) engaged in a particular type of what (activity), here-and-now, than you have pulled of a Discourse. Whatever you have done must be similar enough to other performances to be recognizable. However, if it is different enough from what has gone before, but still recognizable, it can simultaneously change and transform Discourses.
>
> (Gee 2005, p. 27)

In other words, the 'activity' (the 'doing') is the key to being recognized as a particular kind of person, whether it has to do with enacting the Discourse of physics or the Discourse of a certain kind of femininity. Thus, there are important parallels between Gee's Discourse analysis and situated learning theory (or post-structural gender theory for that matter). In fact situated learning theory is acknowledged as a major source of inspiration for Gee's theoretical modelling. All in all, Gee's approach to discourse analysis is one constructed from numerous different sources of inspiration; what makes it 'his' Discourse analysis is the particular way in which he has 'adapted and mixed together the ingredients and, thereby, made the soup' (Gee 2005, p. 6). Gee's book is not 'only' a book about research methods, but as he points out, a method always goes with a theory and consequently his book offers a theory about 'the nature of language-in-use'. However, he also points out that the tools and strategies presented are meant to be flexible enough for a variety of applications; one may use his theoretical modelling as a whole, but it is also possible to choose to use particular parts:

> This book will introduce various tools of inquiry … and strategies for using them. … But the reader should keep in mind that these tools of inquiry are not meant to be rigid definitions. Rather, they are meant to be 'thinking devices' that guide inquiry in regard to specific sorts of data and specific sorts of issues and questions.
>
> (Gee 2005, p. 7)

I have chosen to make use of one specific tool of inquiry from Gee's theoretical modelling, namely that of Discourse models, thereby following Gee's own advice that:

> …this book is meant to 'lend' readers certain tools of inquiry, fully anticipating that these tools will be transformed, or even abandoned, as readers *invent their own versions of them or meld them with other tools* embedded in different perspectives.
>
> (Gee 2005 p. 5, emphasis added)

Gee (2005) describes Discourse model's as '"theories" (storylines, images, explanatory frameworks) that people hold, often unconsciously, and use to make sense of the world and their experience in it' (Gee 2005, p. 61). As with any other kind of 'theories' Discourse models are simplifications of reality that 'attempt to capture some main elements and background subtleties' (Gee 2005, p. 61). Thus, the Discourse models can help us to understand a complex reality by focusing on some central things and leaving out the details. In doing so the Discourse models embed assumptions about what is 'appropriate', 'typical' or 'normal' (in terms of, for example, attitudes, ways of acting or talking), but also what is seen as marginal or not 'appropriate, 'typical' or 'normal'. Further, it is not uncommon for a Discourse model to be signalled by metaphors (Gee 2005).

We learn the Discourse models from our experiences, but as our experiences are shaped and normed by the social and cultural groups we belong to (Gee 2005), it is likely that Discourse models will to some extent be shared among the members of a community of practice. I would therefore argue that Discourse models can be understood as constituting an important part of the shared practice of a community. This is also the reason why Gee chose to use the term Discourse models (as opposed to, for example, cultural models) as these 'theories' are connected to specific Discourses, in other words 'specific socially and culturally distinctive identities people can take on in society' (Gee 2005, p. 61).

In much of his discussion about Discourse models Gee treats them as real entities, as something existing in the minds of people and in texts and social practices. The reason for this, he explains, is primarily to make the concept less abstract and easier to grasp. However, principally he is interested in Discourse models as 'tools of inquiry' (Gee 2005) and that is also how I will use the concept; as a way for me, the analyst, to talk about and, thus, from my perspective constitute the world of the interviewees. When confronted with an actual piece of talk, text or similar (in my case the interview transcripts), the 'tool' Discourse model can lead to questions such as:

- What Discourse models are relevant here? What must I, as an analyst, assume that people feel, value, and believe, consciously or not, in order to talk (write), act, and/or interact this way?
- What sorts of Discourse models, if any, are being used here to make value judgments about oneself or others?
- How consistent are the relevant Discourse models here? Are there competing or conflicting Discourse models at play?

(Gee 2005, p. 92-93)

An example of a Discourse model is the introspective Discourse model developed by Johannsen (2007) in his study about attrition in university physics. In their talk about why they left their physics education the students in Johannsen's study were found to look inwards, to interpret their experienced difficulties as something about themselves (for example, that they were not good enough), not something about the physics education – which Johannsen characterised in terms of an introspective Discourse model.

5.3 Positioning

Situated learning theory does provide useful tools for analysing the physics student community of practice, but, in line with previous critique, I would argue that it lacks depth when analysing the individual's relation to the community of practice. Previously, there have been two principal types of critique of situated learning theory; that the concept is too localized and relations to external structures are largely ignored (see, for example, Nespor 1994 and Davies 2005), and that the internal structures of communities of practice have not been theorized enough in terms of power and hierarchy (see section 3.1.2). I would argue that these types of critique in fact can be seen as different sides of the same coin; namely that the individuals' participation, as affected both by internal and external structures, have not been enough theorized within situated learning theory. Following Linehan and McCarthy (2000) I further argue that positioning theory can serve as a powerful analytical complement to situated learning theory for doing this more fine-grained analysis of the individual and their participation in a community of practice. They further argue that the concept of positioning can be used to 'point more clearly to the production of persons and practices in concrete interactions than is apparent in the communities of practice approach' (p. 442). I agree with this and in particular see positioning as a powerful analytical tool when analysing how practice and identity are negotiated by the participants in the interview situation.

Positioning theory is a framework with its roots in social psychology and examines how participants position themselves relative one another in conversations (van Langenhove and Harré 1999). According to van Langenhove and Harré (1999) 'positioning can be understood as the discursive construction of personal stories that make a person's actions intelligible and relatively determinant as social acts and within which the members of the conversations have specific locations' (p. 16). In a conversation persons can both position themselves (reflexive positioning) and position others (interactive positioning) as they produce joint storylines. In particular the concept of positioning focuses attention to the dynamic of individuals' participation in a conversation in contrast to the more static concept of 'role' (Davies and Harré 1999). As a further elaboration on positioning theory, Harré and van Langenhove (1999) write:

> Positioning theory focuses on understanding how psychological phenomena are produced in discourse. Its starting point is the idea that the flow of everyday life, in which we all take part, is fragmented through discourse into distinct episodes that constitute the basic elements of both our biographies and of the social world. The skills that people have to talk are not only based on capacities to produce words and sentences but equally on capacities to follow rules that shape the episodes of social life. Not only what we do but also what we can do is restricted by the rights, duties and obligations we acquire, as-

> sume or which are imposed upon us in the concrete social context of everyday life.
>
> (Harré and van Langenhove 1999, p. 4)

In other words, the social context – here the physicist community of practice – gives the participants a sense of what they 'ought to do', but is not determining of their actions. Thus, 'participants position themselves in relation to one and another by drawing on the possibilities within a community of practice' (Linehan and McCarthy 2000, p. 442). Further, within a community of practice there are, as argued by Linehan and McCarthy, many, often contradictory, options created by and available to the participants. By exploring what positionings are available for the students, I am able to characterise the practices of a physics student community of practice, as outlined below.

In principal positioning theory was developed for the analysis of how people position themselves and others in a conversation, and strictly applied to an interview situation this would imply a focus on how the interviewer and the interviewee position themselves and each other. An example of this from my empirical material is when the graduate student Tor interactively positions me, the interviewer, as an experimental physicist:

> Tor: It's nice that you're in experimental physics, then I can take examples...

However, my main focus is not on the positionings taking place between the interviewer and the interviewee, but rather how the interviewees talks about their positionings of themselves and others 'outside' the immediate interview situation. To illustrate this I use an excerpt from the interview with Hanna, in which she can be interpreted as firstly positioning herself (reflexive positioning) as a non-physicist, and secondly as positioning the other people in her research group as physicists (interactive positioning):

I:	You're not doing the most experimental project of the people I've interviewed, so that question is a bit... [we are both laughing]. Do you see yourself as a physicist then?
Hanna:	No, not really that either [laughter]. Actually. I've never thought about it like that.
I:	Why not?
Hanna:	... Well ... It may also be the education I've chosen, that then it's not really the same as ... as, for example, Engineering Physics.
I:	But the other people in your research group then, are they physicists?
Hanna:	Yes...

Thus, I apply the analytical tool of positioning in quite a loose sense. In particular, the positioning of a non-present other is quite different from the interactive positioning of someone participating in a conversation.

Previously, the use of positioning theory within the realm of science education research has been quite limited, some examples of such research are Barnes (2004), Ritchie (2002) and Berge and Danielsson (2008).

5.4 Concluding Remarks

In this chapter the third and last part of my conceptual framework – the analytical tools – have been introduced. These tools serve as the bridge between my empirical material and my theoretical framework in the sense that they complement this framework with analytical entrance points to the interview transcripts.

In the next chapter I discuss methodological considerations of my research, such as issues of trustworthiness and ethics. I also describe how the empirical data was collected and analysed.

CHAPTER 6

Methodological considerations

6.1 Introduction

In the three previous chapters I presented the conceptual framework of my research, but apart from the choice of appropriate theories there are also additional methodological considerations to take into account when doing research, for example, how to collect the empirical material and how to establish trustworthiness. Such methodological considerations are the focus of this chapter.

I begin the chapter with a discussion about how the empirical data was collected and the reflections involved in the collection and treatment of the data. Next, I focus on issues of trustworthiness, specifically how I dealt with them in my research. I also briefly discuss ethical considerations. Lastly, I outline how the analysis was carried out in practice. Due to the explorative character of my research this last part is crucially important in order for the reader to follow (and critically evaluate) the process of my research. However, the explorative character also means that it is difficult to do justice to the complexity of the long and winding research process in a written account. As such, the aim of this entire chapter (not just the last section) is to introduce the reader to various aspects of the research process; choices I was faced with and decisions made.

6.2 Data Collection

In order to empirically explore physics students' simultaneous constitution of the practice of physics and identities as physics students/physicists, I chose to make use of individual, qualitative research interviews. The main reason I decided to use interviews and their transcripts as my data – instead of, for example, questionnaires – is that interviews, especially when they are done in a semi-structured fashion, facilitate a much deeper probing into the experiences of the interviewees (Kvale 1996). Secondly, it was important for me to let the students themselves give voice to their experiences. For this reason I chose not to do (participant) observations, since this would have implied that I was making interpretations about the students rather than letting them give their own rich interpretations. Thirdly, semi-structured interviews also have the advantage of giving access not only to what the interviewees actually are saying, but also to what the interviewees find easy to talk about, and even what they struggle to express. While observations could have been used as a complement to the interviews, for example, to assist in constructing relevant interview protocols, I did not see them as necessary from this perspective; my own experiences as a physics student (and physics teacher) have, at least to some extent, filled this purpose. Of course, own experiences of this kind can never be reflective in the way that fieldwork can, but personal experiences can still contribute with important insights thereby raising the quality of the interviews.

The number of students enrolled in the undergraduate Master of Science Programme at the University is quite limited, and the number of graduate students willing to participate in my study was also very limited. This led me to choose individual interviews over focus group interviews. Firstly because, the limited number of possible interviewees simply made it difficult to collect enough interviewees to put into focus groups of a viable size. Secondly, because the limited number of students enrolled in physics also meant that the undergraduate students interviewed were likely to know each other.

While the focus group interview can have the advantage of creating a more relaxed discussion by letting the students support and encourage each other, it can also be difficult for the interviewees to talk freely, especially if they know each other. Thus, the individual interview has the advantage of letting the interviewee talk about potentially sensitive issues, involving, for example, teachers and other students, in a way they might not be comfortable with if other students are present. In focus groups there are also complex group dynamics and power hierarchies that affect who speaks and what can be said, hence certain views and certain people might dominate. But, on the other hand, the group may also raise the consciousness of the participants and be empowering (Robson 2002). In summary, I concluded that the disadvantages with using focus groups outweighed the advantages for my research purposes, and decided to use individual interviews.

6.2.1 The Qualitative Research Interview

Kvale (1996) describes the qualitative research interview as a conversation where two people talk about a subject both are interested in and thereby build knowledge. The aim of such an interview is to understand aspects of the life-world of the interviewee from the interviewee's own perspective. Hence, the qualitative research interview is typically semi-structured, and this is the type of interview I have used in this study (see Appendices C and D). A semi-structured interview, Kvale (1996) writes, is neither a totally open conversation nor one directed by a detailed questionnaire. Instead an interview protocol, where certain themes and questions are given, is used to guide the conversation from the interviewer's perspective. Throughout the interview follow-up questions are used in order for the interviewer to understand the meaning of what the interviewee articulates, and to ensure that she has correctly interpreted what has been said. This is however not the same as trying to get unambiguous answers, since:

> The aim of the qualitative research interview is not to end up with unequivocal and quantifiable meanings on the themes in focus. What matters is rather to describe precisely the possibly ambiguous and contradictory meanings expressed by the interviewee.
>
> (Kvale 1996, p. 34)

Thus, in a qualitative research interview it is central to access as many nuanced descriptions as possible in order to capture as much diversity as possible. In order to achieve this, I as the interviewer must further strive to create an atmosphere that allows the interviewees to talk as freely and honestly as possible. The strength of such interviews (as compared to, for example, a questionnaire) lies in the depth to which different issues can be explored and the openness for the discussion to take routes other than the ones perhaps anticipated by the interviewer. In this context it is also important to remember that the interview situation is co-constructed by the interviewer and the interviewee; thus, the interviewer also plays an active and unavoidable part in what contextual identity the interviewee constitutes (Gee 2005). I have used an excerpt from the interview with the Master's student Hanna to illustrate a moment of very explicit 'co-construction'. This interview excerpt is also an example of how the interview situation can provide an opportunity for reflection and make the interviewee see something from a new perspective. Here Hanna is talking about how 'simple' her Master's research project is, how it is something anyone could do, but confronted with my question about whether she would had been able to do it earlier in her education she changes her mind:

I:	Do you experience ownership over your project? Over what you do?
Hanna:	... Erm ... Maybe a bit more now towards the end.
I:	Yes.
Hanna:	When one starts to realize that it's me who has done this and ... there is no one else that knows what I've done, almost. So, then you get to do it…
I:	What has changed during the course of the Master's research project then?
Hanna:	Mmm… It might be that to begin with I thought that what I'm doing is so very simple and…

I: What do you mean by that?

Hanna: Well, easy. I sit here and simulate small things.

I: Aha

Hanna: And anyone could do that and it doesn't take much to do that. And, I also felt that I'm not doing much, I've been here for so long and I haven't done much, somehow. But then, and that might be because I come straight from school, where everything is supposed to go so quick and ... then you start to realize that maybe it's like, I've been working all the time, so it's not like I'm wasting my time...

I: No, no.

Hanna: ...maybe it doesn't go faster than this. And, ok, it might not be the most difficult thinks I've done, but ... the conclusions and so on one has made from it, it's still, are valuable after all.

I: But could you have done it earlier in your education? You say that it's very simple.

Hanna: Erm... Maybe... But it's still that you have matured a bit more and do something independent after all and... still, it takes this knowledge about the subject, and those I didn't have before, so that I couldn't, no... No, actually I don't think so. But maybe I would [inaudible] but it would have been very difficult, then you would have had to read a lot of stuff.

6.2.2 My interviews

At the start of all the interviews I introduced myself as a PhD student in physics, explaining that I am doing physics education research, but consciously stressing my background in physics. For example, I explained to the interviewees that my interest in laboratory work stems from my own background in experimental physics. By such an explicit positioning of myself as a physicist I constructed the interview situation itself as a physicist community, as two physicists talking about the learning and doing of physics. Consequently, this allowed me to explore how the interviewees constitute their identities as physicists in relation to a physicist community as opposed to in relation to, for example, the general public or an interviewer from a different discipline. The construction of the interview situation as a physicist community thus contributed to a more nuanced exploration of how the students constitute their identities as physicists, and made their sharing of experiences related to the procedures and content of physics more meaningful. I also shared my own general experiences of being a physics student in order to establish a trusting relationship with the interviewees.

In some senses, my own experiences from physics can be compared to participant observations, in that they not only gives insight into a particular phenomenon but also contribute to building a trusting relationship with the participants. In the case of the PhD student Tor it was very explicit how this 'shared insider-perspective' did in fact influence the interview situation. For example, he described himself as a 'physics nerd', an expression he claimed he only would use among fellow physicists and, further, he also explicitly acknowledged that the fact the I have a background in experimental physics allowed him to use concrete examples from that realm. Still, I am always mindful that unavoidable power relations can emerge between the interviewer and the interviewee, which can have a significant influence on guiding the interview situation. Moreover, there is also the possibility that my perception of a shared insider-perspective contributes to a feeling of 'knowing what the students are talking about' and that I therefore might not ask them enough follow-up questions. This illustrates how it, in a sense, is the researcher that is the key instrument in qualitative research (Bogdan and Biklen 1996), not only in the analysis but also in the data collection. This was also made apparent when several of the woman students interviewed explicitly identified with me as a fellow woman physicist in referring to, for example, what they perceive as shared experiences.

The interviews with the undergraduate students

The undergraduate students[16] were interviewed in the spring of 2005. All the interviewees were at the time enrolled in an undergraduate physics degree programme at the University. I recruited the interviewees by visiting lectures and asking for volunteers. The selection of students was strategic; I wanted students from different years and ages and was aiming for a relatively even distribution of men and women. The selection was not meant to be representative, but rather to reflect a diversity of students and possible identity constitutions.

The interviews lasted between 25 and 70 minutes. I audio-recorded the interviews and then transcribed them verbatim. All the undergraduate students interviewed were native Swedish speakers, and consequently all the interviews were conducted in Swedish. The interviews were semi-structured, following the protocol found in Appendix C. The interview protocol was developed with a broad focus on issues regarding the students' experiences of doing laboratory work in physics. Taking situated learning theory as my theoretical starting

Pseudonym	**Year**
Ann	3rd year
Susan	3rd year
Lina	1st year
Bo	2nd year
David	2nd year
Paul	3rd year
Dan	1st year
Ida	1st year
Lars	1st year
John	1st year
Lisa	1st year
Mia	1st year
Kalle[15]	4th year

Table 2. List of the pseudonyms used for the undergraduate students, and their year of study.

[15] Kalle had at the time of the interview just began his Master's research project, but was interviewed about his experiences of the student laboratory.

[16] Students enrolled in the Master of Science programme ('Naturvetarprogrammet') prior to the Master's research project.

point the interviews were particularly focused on issues related to identity constitution and perceived norms of the physicist community. However, it should be noted that at the time of the interviews I had not, as yet, developed my theoretical framework. The initial development of this only took place during the analysis of the interviews. Thus, the interview protocol was not specifically aimed at themes brought to the fore in the theoretical framework development. In short, the students were asked to talk about their experiences of doing laboratory work in physics; what they saw as valuable skills to have and to develop for the student laboratory, what previous experiences they perceived as being useful, how they thought of themselves as laboratory students, the purpose of laboratory work etcetera. Then, during the final stages of the interview issues of gender in physics education were explicitly brought up.

The interviews with the graduate students

The interviews with the graduate students[17] were carried out in 2007. To recruit interviewees I sent out an email asking for volunteers who were currently doing their Master's research project in experimental physics.

Pseudonym	
Tor	PhD student
Marina	Master's student
Ann	PhD student
Cecilia	PhD student
Hanna	Master's student
Erik	Master's student
Klara	Master's student
Karin	PhD student
Martin	Master's student

Table 3. List of the pseudonyms used for the interviewed graduate students.

The Master's students interviewed all contacted me in response to this email (some were recommended to contact me by their supervisors). I also sent out a similar enquiry to the PhD students at a physics department, which provided me with two more interviewees. Further, I sent a personal email to Ann, and she in turn recommended Karin. Thus, the participants were self-selected, but the selection was strategic in terms of covering both Master's students and PhD students, both men and women and students with different educational backgrounds. However, being self-selected they should not to be taken as a representative sample. The original aim was also to capture a greater diversity in terms of, for example, men/women, educational background and age. In the end it turned out to be quite difficult to find interviewees and so I ended up interviewing all students who volunteered.

All interviews, apart from the one with Tor, were conducted in Swedish. Tor was interviewed in English, his second language. Marina and Martin were also asked if they wanted to be interviewed in English, Swedish being

[17] Students currently working with their Master's or PhD projects.

their second language, but both of them preferred Swedish. The interviews lasted between 40 and 60 minutes, they were audio-recorded and transcribed verbatim. The interviews were semi-structured, following the protocol that can be found in Appendix D. At the time of this second set of interviews I had already developed major parts of the theoretical framework and was interested in exploring issues related to the further development of this framework, something that influenced the development of the interview protocol. Consequently, the students were asked about their progression from physics student to physicist, differences and similarities between the student laboratory and the research laboratory, whether they viewed themselves as physicists, norms of the physicist community, and how they learnt to become physicists. At this point 'identity', in particular in regard to gender, had emerged as an even more explicit focus of my research. This explains why more questions about physics and gender were included as well as more detailed biographical questions about, for example, the parents' occupations.

6.2.3 Transcription and Translation

The transcription of the interviews may seem like a fairly straight forward thing to do, but as pointed out by Kvale (1996) transcription is not just a matter of converting spoken language into writing:

> Transcribing involves translating from an oral language, with its own set of rules, to a written language with another set of rules. Transcripts are not copies or representations of some original reality, they are interpretative constructions that are useful tools for given purposes.
>
> (Kvale 1996, p. 165).

Consequently, it is important to keep in mind that the transcription is always an interpretation, no matter how much detail is strived for. Furthermore, different research interests call for different approaches to transcription; a psychological or detailed linguistic analysis needs a transcript that includes pauses, repetitions and tone of voice (Kvale 1996). In my analysis I am primarily interested in what the students are saying, and only to a lesser extent how they are saying it. Consequently, I have chosen to transcribe the interviews verbatim, but with little additional detail; longer pauses have been marked with [...] and obvious emotional expressions such as laughter have also been marked. I transcribed all the interviews myself, which not only gave me control over how the transcription was made (in terms of detail, for example), but also gave me the opportunity for intense engagement with the empirical material. Once transcribed, I went through the transcriptions while listening to the audio-recordings, correcting the mistakes found.

During the analysis phase I worked with the original Swedish (and in one case, English) transcripts, returning to the audio-recordings when this was judged to be necessary. The interview excerpts chosen to be included in the thesis were then translated, by me, with the aim of capturing the meaning of the utterances rather than being literal translations. When there were doubts regarding how to translate a certain utterance I discussed the translation with my colleagues. In the translations I have chosen to remove repetitions, stuttering and some colloquial speech, to enhance the readability. The English transcript was edited in the same way.

Vignette: The Interview with Karin

In the following I give a description of what could be considered 'a typical interview', in order to share a feeling for how the interviews were conducted.

After the interview with Ann I asked her if she had any PhD-student colleagues that might be interested in being interviewed for my research project. She then told me about Karin, another recently employed PhD student in her research group. Further, Ann told me a little about Karin's research, and Karin really seemed very suitable for my research project. I asked Ann for Karin's email address so that I could contact her and Ann also informed her colleague that I would be in touch with her. A couple of days later I emailed Karin and asked her if she would be willing to participate in an interview. She said yes, we agreed on a time and decided to meet outside my office. When we met for the interview I introduced myself briefly and offered that we could go and get a cup of coffee or so; an offer she turned down. I have generally found that the mundane activity of a walk to the coffee machine and the associated small talk is a good way to get the interviewee to relax and start talking. The actual interview with Karin took place in a combined office space and seminar room. The room is furnished with both a large table with chairs and a smaller table with a couch and an armchair. I let Karin decide where we should sit; she chose to sit down on the couch and I sat down in the armchair, placed at right angles to the couch. I began the interview with a brief explanation of my research process, how I am interested in laboratory work and in issues of gender and physics, motivated by my own background in experimental physics. I also told her that the interview was all about her thoughts and experiences and that I hoped for it to be as much of a conversation as possible, as I did not have a well-defined set of questions but rather an interview protocol highlighting a number of themes I wanted to discuss. Finally I asked her if it would be OK if I recorded the interview, explaining that I would be the only one listening to the actual recording, but that the transcript might be shared with my colleagues, and that her participation in my study, of course, would be totally confidential and anonymous. She agreed without hesitation or further questions. As with many of the interviewees she was not particularly interested in the finer details concerning her participation in the study, as it, according to her, did not concern very personal matters. To get the interview going I started with some questions about biographical background data (where she grew up etc.) and then moved on to questions about her PhD project. The rest of the interview was guided by the interview protocol, but rather than following it linearly I let the different themes emerge naturally; guided by Karin's answers – which meant that I was jumping around quite a bit in the protocol. The entire interview lasted for about one hour and I had the impression that Karin was rather relaxed and felt comfortable in the interview situation. I rounded off the interview by asking Karin if she had something she wanted to add, which she did not, and told her that she was welcome to contact me at any point, had she got any questions about the study or her own participation in it.

6.3 Trustworthiness

Qualitative research in general can be seen as being concerned with knowledge as a human construction, and the objective is thus not to reach the 'true reality' but rather to understand the lived experiences as told by, for example, the interviewees. Commonly, in positivist research, issues of quality are discussed in terms of validity, reliability and generalizability. However, within qualitative research – where the researcher seeks understanding, not 'facts' – alternative constructs are needed, such as, credibility, dependability, transferability, and confirmability (Lincoln and Guba 1985; Guba and Lincoln 1989).

6.3.1 Credibility

Credibility parallels the criteria of internal validity, but

> instead of focusing on a presumed 'real' reality, 'out there', the focus has moved to establishing the match between the constructed realities of respondents and those realities as represented by the [researcher] and attributed to various [respondents]
>
> (Guba and Lincoln 1989, p. 237)

In other words, the key is to be 'true' to the storied experiences of, in my case, the interviewee. To ensure that my story is in fact their story the continuous discussions of my findings with my research colleagues were of crucial importance. 'Prolonged engagement' with the site of inquiry is commonly used as a way to ensure credibility; even though I have not participated in the laboratory work together with the interviewees, I do see my extensive experience of studying physics as providing an essential component of this credibility.

6.3.2 Dependability

In research aimed at predictive ability it is of vital importance to make sure that the data is stable over time, that it is reliable. In qualitative research such as mine, on the other hand, changes must be expected; this is seen as a desirable part of the evolution of the research process. As the research is maturing it is likely that methodological decisions will shift the hypotheses, constructs etc. and perhaps even the research purpose. These shifts, however, need to be tracked by the researcher (and trackable by others) – which is what ensures dependability (Guba and Lincoln 1989). In the case of my research, I kept a research journal in order to document the progression of my research, for example, how the research purpose has evolved.

6.3.3 Transferability

My research is very much situated in the context where it took place, as is most qualitative research, and consequently the results are not expected to be generalizeable in the 'predictive research' sense of the word, but in the 'understanding research' sense of the word. Instead, I have strived for transferability to 'provide as complete a data base as humanly possible in order to facilitate transferability judgements on the part of others who may wish to apply the study to their own situation' (Guba and Lincoln 1989, p. 242). Thus, I will provide an account of the research from which the reader can create a deepened understanding of some aspect of their life, that they themselves can 'generalize' from. This can, for example, take the form of readers raising questions about their own practice and their own ways of knowing (Clandinin 1992). An analogy is how reading an autobiography can help to shed new light on issues from life for the reader; whether it has to do with recognizing ourselves in the text or from gaining a different perspective.

Transferability has also been characterised in terms of 'naturalistic generalization' (Stake 1994), where readers make their own 'generalizations' based on what *they* find in the study. Thus, for quality assurance, my responsibility as a researcher is to provide readers with a sufficiently 'thick description' (Lincoln and Guba 1985, p. 316) to enable the readers themselves to relate to my interpretations. Here it is important to point out that whether readers agree with my interpretations or not is not the condition for a transferable study. It is the readers' ability to understand and thereby judge my interpretations that is necessary.

I have tried to ensure transferability by providing such a sufficiently 'thick description' including, for example, generous interview excerpts in Part III of the thesis, thereby making it possible for the readers to judge whether they agree with my interpretations or not. I also give a detailed account of the conceptual framework. For this purpose I have also included lengthier excerpts from two interviews (one with an undergraduate and one with a graduate student) in Appendix E.

6.3.4 Confirmability

Confirmability can be thought of as a metaphor of objectivity, in that both are concerned with 'assuring that data, interpretations, and outcomes of inquiries are rooted in contexts apart from the [researcher] and are not simply fragments of the [researcher's] imagination' (Lincoln and Guba 1989, p. 243). Confirmability is assured by seeing that the data can be tracked to its sources and that in the thesis it is made explicit how the outcomes has been logically assembled from the data.

Here, just as in the case of transferability, a suitably thick description will play a crucial role. By providing lengthy excerpts from the interviews together with full details of the process I have made it possible for readers to trace my analysis back to the empirical data. Throughout the research process I have also continuously shared my findings, conclusions and tentative analysis with my research colleagues – in order to ensure that my interpretations are plausible, thereby giving the research process confirmability. Primarily this sharing has taken place within my research group, both informally and in formal seminars where we have read and discussed my interpretations. I have also discussed my interpretations with researchers outside the research group and presented earlier versions of my analysis at a number of international conferences. This ongoing process of sharing and discussion has given me much valuable input and at times it has caused me to rethink critical aspects of my analysis.

6.4 Ethical Considerations

In planning, carrying out, and reporting on my research I have taken various measures in order to follow the research ethics principles in humanistic-social scientific research developed by The Swedish Council for Research in the Humanities and the Social Sciences (Vetenskapsrådet 2002). These measures are outlined in the following:

I began all the interviews by introducing myself and gave a brief introduction to my research project. I also explained to the interviewee that their participation in the study was voluntary and that they could withdraw from the interview at any point or choose not to answer certain questions. I also asked permission to audio-record the interview. All interviewees agreed both to participation in the study and to the audio-recording. Further, I informed the interviewees that I would guarantee them confidentiality and anonymity. After the interview the interviewees were given the opportunity to get more information about the research project and I also told them that they could contact me at any point, had they any questions. They were also informed about how I would work with the audio-recording and that the results of the study would be published in articles and my PhD thesis.

In order to preserve the anonymity of the interviewees, all names used in the thesis are pseudonyms; this applies both to the interviewees themselves and other people mentioned in the interviews, such as other students, supervisors and teachers. In the case of these subsidiary characters, several different pseudonyms may have been used for the same person. The graduate students were given the opportunity to choose their own pseudonyms, but only one student did so. In the biographical descriptions I have excluded information regarding home town, universities attended and similar information that may expose the identity of the interviewee. In order to make the identification of students extremely unlikely I have also changed some other biographical information that I did not judge to be relevant for the analysis, such as number of brothers and sisters. The empirical material that was shared with research colleagues did not include any information about identifiable individuals.

So far I have discussed issues of trustworthiness and ethics independently of each other, but these aspects are also intertwined. Thus, the issues of trustworthiness discussed earlier also have an ethical component; to be true to the empirical data and represent it in a way that gives justice to the lived experience of the interviewees – what Blumenfeld-Jones (1995) calls fidelity. But not only must the researcher maintain fidelity 'towards the story of a person (what the person makes of his or her story) [but also] towards what that person is unable to articulate about the story and its meaning (and the context in which the story exists)' (Blumenfeld-Jones 1995, p. 28). The notion of fidelity also brings to the fore how research such as mine contains several layers of interpretation and that the final research account involves layers of intention and reconstruction, both from my perspective and from the perspective of the interviewees.

6.5 Analysis in Practice

In the introduction to Part II I described how I step-wise developed the conceptual framework of my research, starting from a broad theoretical staging of situated learning theory that then was sharpened into my theoretical framework and finally complemented with analytical tools – always working with a close interaction between the empirical material and the theories employed. Once I had completed the conceptual framework I analysed the entire empirical material (in the case of the interviews with the undergraduate students, reanalysed) and it is this analytical process that I will primarily focus on in this section. But first a few words about the initial stages of analysis.

In a sense the analysis of the empirical data began already during the collection of it, during the interview itself, in that I, the interviewer, interpreted the meanings of the interviewee's utterances and asked for the clarifications I found necessary. Consequently, the audio-recorded interviews by all means and purposes a co-construct by me and the interviewee. I then transcribed the interviews myself and it is these verbatim transcripts that are the basis for the analysis. In the early stages of the analysis, while transcribing and later reading the interviews repeatedly, I noted down common themes. In the repeated readings of the transcripts these themes were further explored and refined. At this stage the analysis was guided both by key aspects from the emerging research purpose (such as gender and laboratory work) as well as by themes surfacing in the empirical material (such as skills and experiences viewed as important by the interviewees). Concretely, I worked with making notes and underlinings in the transcripts. I also created mind-maps (organised around the individual students as well as around emerging themes), wrote summaries of the interviews and categorised emerging themes in excel-sheets. Having started the analytical process working closely with the interview transcripts, contrasting and comparing utterances within and across transcripts, I increasingly involved situated learning theory, which included drawing on various related studies. Eventually this analytical process, the interplay between empirical material, previous research and situated learning theory, led me to construct an initial version of the theoretical framework. The emerging theoretical framework was then refined as the analysis progressed, through a close interaction between the empirical and the theoretical. The analytical outcomes were also continuously discussed with research colleagues, at seminars and conferences. Eventually, the theoretical framework together with initial empirical results were presented in my licentiate thesis (Danielsson 2007). Next, I conducted my second round of interviews. In working with the analysis of the transcripts of these interviews it became increasingly apparent that the theoretical framework needed to be complemented with analytical tools to form a bridge between the empirical material and the theoretical framework. For this purpose I included the tools of Discourse models and positioning into my conceptual framework (see Chapter 5).

The final analysis was carried out in three subsequent steps of increasing levels of abstraction, which are given in Chapters 7, 8 and 9.

In the first stage of the analysis I kept closely to the empirical material. By employing the analytical tool of positioning to the interview transcripts I here made an initial exploration of how the students constitute their identities as physicists and constitutes the practice of physics. An important theme at this stage of the analysis was the students' approaches to laboratory work and what they perceived as being desirable approaches within physics. Another theme was the students' constitutions of the boundaries of physics. The final outcome of this stage of the analysis is the student narratives found in Chapter 7.

In the second stage of the analysis, the focus was shifted from the individual students to how they as a collective group can be understood as constituting the practice of physics. The exploration of the practice perspective was focused on three major themes:

1) The Discourse models the students use to make sense of the practice of physics as enacted in the student laboratory and the research laboratory respectively.

2) How the students characterise ‘the physicist’, as a means to identify some of the norms the students perceive as defining the physicist community.
3) How the students perceive the boundaries of the physicist community, with a particular focus on the relation between the student community and the community of professional physicists.

The Discourse models are grounded in the student narratives from Chapter 7, but in constructing them I also worked with all the interview transcripts – moving between the student narratives, the interview transcripts and the emerging Discourse models in an iterative fashion. The construction of the Discourse models is further described in section 8.2. In connection with my construction of the Discourse models I also worked with themes (2) and (3) simultaneously refining and exploring these themes. The outcome of this stage of the analysis is presented in sections 8.2 to 8.4. Finally, the theoretical framework was brought into the analysis for a deepened analysis of the outcomes presented in sections 8.2 to 8.4. Here I, in particular, focus the analysis on the gendering of the physicist community of practice.

It should also be mentioned that while the practice perspective (the primary focus of this second analytical stage) and the identity perspective (the primary focus of analytical stages one and three) can be thought of as illuminating different aspects of a community of practice, there is no absolute distinction between the two; the practice does not exist independently of the practitioners, it is their doing that is the practice, and vice versa, the practice provides a repertoire for the identity constitution. The boundary between the analysis of the practice and the analysis of the students’ identity constitution is therefore necessarily blurred.

In the third and final stage of the analysis I returned to a thematisation that followed the individual students. Here, the narratives from Chapter 7 were combined with the full complexity of the conceptual framework, as well as the outcomes from Chapter 8, to deepen the analysis of how students constitute identities as physicists and how they negotiate the norms and boundaries of the physicist community of practice. For example, in this stage of analysis the Discourse models were used to explore the students’ negotiation of their own and others practice of physics and their related identity constitution. I also looked at how the students positioned themselves and other students/groups of students in relation to the two Discourse models and how the students positioned the Discourse models in relation to the practice of physics (are they central, marginal or ‘outside’).

Here it needs to be pointed out that the three analytical steps were not carried out linearly, but rather in a reciprocal fashion, where I, for example, allowed the third step of the analysis to influence and sharpen the Discourse models. The analysis also continued in the writing process, as expressed by Richardson (1994): ‘I write because I want to find something out, I write in order to learn something I didn't know before I wrote it' (p. 517). Consequently, the writing of this thesis is, in itself, the last step of the analytical process; the writing of the thesis has, for me, not been about transforming a finalized analysis into text, but the writing has in itself been a method of discovery.

The final outcome of my analysis is a mix of thematic accounts and personal narratives, which together aim to:

> ...both understand and try to give an account of the ways in which the individual is shaped by the situation and shapes the situation in the living out of the story and in the storying of the experience.
>
> (Clandinin 1992, p. 128)

Thus, the story I – the researcher – tell is not simply a reproduction of the students' stories. By re-telling their stories through my research perspective, I aimed to create a final story that 'must fit the data while at the same time bring an order and meaningfulness that is not apparent in the data itself' (Polkinghorne 1995, p. 16). The final aim of such analysis is to create a rich and explanatory description of the data.

Finally, I would once again like to stress that my analysis by no means proceeded in a linear manner; from research purpose, to collection of empirical data, to application of analytical tools, to interpretation using the theoretical framework, to the final writing up of the thesis. Rather, all these different steps are complexly intertwined. For example, the theoretical framework most fundamentally guides how I formulated the purpose of my research and it also guides what positionings and Discourse models I chose to focus upon. And, vice versa, the empirical data is also allowed to speak back to the theoretical framework.

6.6 Concluding Remarks

The aim of this chapter has been to shed some light on the research process, choices faced, and decisions made. As such the chapter has included descriptions of how the data was collected and analysed as well as discussions about how trustworthiness was established.

Having now completed Part II of the thesis, I have laid the final piece of the foundation for the presentation of the analytical outcomes of my research which follows in Part III.

PART III

Analytical outcomes

Introduction to Part III

How do physics students simultaneously constitute the practice of physics and their own physicist identities in relation to this practice? And how can these constitutions be understood as gendered? These are the questions that drive the empirical investigation presented in Chapters 7 to 9, and the answers to these questions address the purpose of this thesis as stated in Chapter 1.

In Chapter 7 a selection of the students interviewed is introduced, through individual narratives. By such an initiation of the exploration of how the students constitute the practice of physics and their own physicist identities the reader is also familiarised with the students interviewed. The aim of Chapter 7 is accordingly to give an empirical foundation to Part III; to give an empirically close initiation of my exploration of students' constitution of the practice of physics and of identities as physicists.

In Chapter 8 the students are treated as a collective group. Here the focus is on how they collectively constitute the physicist community of practice, with particular attention being paid to how they characterise the practice of the laboratory. In Chapter 9 I turn back to the individual students for a theoretically deepened exploration of how they constitute their identities as physicists, drawing on the theoretical framework as well as the empirical insights from Chapters 7 and 8. The order of Chapters 7 to 9 follows the succession of the analytical process (as described in section 6.5) and through the chapters I strive for a progressively increasing level of abstraction, moving from narratives in Chapter 7, with their relative closeness to the empirical material, to the application of the full conceptual framework in Chapter 9. This way the reader is guided through the analytical outcomes in such a way that more and more aspects of the conceptual framework are introduced, and the level of interpretation is accordingly raised as Part III progresses, contributing both to the accessibility of the analysis and increasing the possibility for the reader to critically evaluate my interpretations.

In total I interviewed 22 students, 13 undergraduate students and 9 graduate students. In Chapters 7 and 9 thirteen of these student interviews are treated in more depth. These students were chosen because they represent a variation in the ways in which they are constituting the physicist community and/or physicist identities. I also sought as much variation as possible in terms of year of study, age, women/men and in the case of the graduate students, educational background. The analysis in Chapter 8 is founded on the thirteen student accounts in Chapter 7, but the other students interviewed are also included in this analysis, as a complement to contrast and compare.

CHAPTER 7

The Practice of the Physics Laboratory: Individual Perspectives

In this chapter seven of the undergraduate students and six of the graduate students are introduced through individual narratives, that are kept relatively close to the empirical material. In doing so this chapter initiates my exploration of how the students constitute the practice of physics and their own physicist identities, but it also familiarises the reader with the students interviewed and the physicist community.

In this chapter I look at how the students talk about their practice in the laboratory. For example, in terms of what they describe as suitable and unsuitable approaches to laboratory work and what they see themselves as skilled and unskilled at. I also look at how they position themselves and others in relation to the physicist community. In the case of the graduate students I also look at how they characterise the physicist and how they describe the transition from physics student to physicist. But, before moving on to introducing the students I will give a brief introduction to the physics student laboratory.

7.1 The Physics Student Laboratory – An Introduction

> The laboratory offers unique opportunities for students and their teacher to engage in collaborative inquiry and to function as a classroom community of scientists. Such experiences offer students opportunities to consider how to solve problems and develop their understanding. Through collaboration, they can also come to understand the nature of an expert scientific community.
>
> (Hofstein and Lunetta 2003, p. 36)

As the above quote indicates, the student laboratory is a highly complex learning environment, with many, diverse learning goals. To concretise some of the possible learning goals in laboratory work in physics I here exemplify with a course in Wave optics from the University, taken from the course website. This course includes three laboratory exercises (polarisation, geometrical optics and wave optics) and each laboratory exercise is categorised as aiming to fulfil one or more of the goals stated in Table 4. The polarisation and wave optics laboratories are aimed at goals 1, 2 and 4, whereas the geometrical optics laboratory is focused on goal 1.

Code #	Categorisations of the laboratory exercises
1	To connect theory and experiment, in particular: a) the importance of experiments for the discipline of physics b) to connect formulas and mathematical models to experimental observables.
2	To learn and to understand experimental methodology and technical equipment.
3	To learn and to understand how experiments are planned and evaluated a) from estimated sources of uncertainty b) from different designs of measurements
4	To understand the methods and ways of thinking in physics research.
5	To learn scientific argumentation and presentation (written and oral)

Table 4. Learning goals used to categorise laboratory exercises in Wave Optics.

As elaborated on in section 2.2.2 of the Literature Review, the alternative conceptions students bring to physics instruction is a well-researched area, but that the students may also have 'pre-conceptions' regarding the procedural aspects of laboratory work, the method, is a much less researched area. In her research Parsons (1995) has described one aspect of the intuitive methods students bring to the laboratory; how some students may have acquired informal knowledge about methods of science through their prior experiences of 'tinkering'. 'Tinkering' is described by Parsons (1995) as 'informal experiences with common experimental materials (magnets, mirrors, electricity, heat and solar energy) and other instruments' (Parsons 1995, p. 204). By surveys and interviews as well as observations of students working with an electricity exercise, Parsons (1995) was able to develop a model of tinkering. In this model the students' purposes for tinkering were divided into four categories:

1. Utilitarian tinkering (useful)
   - is interested in fixing things and often someone who engages in this type of tinkering is usually regarded as the fix-it type,
   - is interested in fixing mechanical and electrical devices because it is perceived as useful for everyday living,
   - looks at tinkering as a means of saving or earning money.

2. Technological tinkering (application)
   - is caught up in the electrical and electronic hardware (the application of knowledge of electricity to electrical and electronic hardware) and is usually fascinated by any electrical or electronic device,
   - is very adaptable to new technological devices,
   - is interested in being in control.

3. Scientific tinkering (theory)
   - is interested in the how and why of electrical and electronic devices or the theory behind how an electrical or electronic device works,
   - is more interested in understanding than doing and will engage in activities for the sole purpose of understanding.

4. Pragmatic tinkering (performance)
   - the emphasis is on performance for grades or doing what the teacher wanted, only tinkers when absolutely required for a well defined task such as school work, or to meet some very limited specific need.

(Parsons 1995, p. 206-207)

Taking the complexity of the student laboratory learning environment as well as the intuitive methods students bring to the laboratory, as described by Parsons (1995), into account it is clear that the student laboratory opens up a wide range of possible identity constitutions. These identity constitutions will be explored in the next section.

7.2 The Undergraduate Students

In the following sections I focus on the individual students' approaches to laboratory work, as well as which approaches they perceive as being the desired ones within physics. I also look at the students' positionings of themselves and others. The analytical tool of positioning focuses attention onto how the students localise themselves and others in relation to the physicist community of practice; how they negotiate the practice of physics and their own identities as physicists. In this way positioning serves as an analytical entrance to the empirical material, one that allows for the more fine-grained analysis of the individual and their participation in a community of practice. Thus, the analysis is here kept close to the empirical material, and apart from laying the foundation for the further analysis the student narratives also serve as a way for the reader to get acquainted with the students and the research context. Each student narrative is started by giving a short biographical note, to 'thicken the description', thereby enhancing the transferability of the results.

7.2.1 Kalle

Kalle was at the time of our interview studying the last year of the undergraduate Master of Science programme. He had just begun his Master's research project in experimental physics, but was primarily interviewed about his experiences of the student laboratory. The Master of Science programme was not his first choice, but his grades from secondary school were not good enough for him to be accepted into one of his preferred engineering programmes. Kalle describes himself as not having been very studious in secondary school; he had activities outside school that took up a lot of his time. Since he did not have the time to study enough he was advised to change from the science stream, which he also, to rather study a vocational programme, focused on industrial work. He also worked at various industrial companies during the summer vacations.

On repeated occasions during the interview Kalle talks about himself as an 'industry worker' and also sees many connections between working in a workshop and working as an experimental physicist. In the interview excerpt below it can also be noticed how Kalle is positioning himself as a 'practical person', but that he perceives that his practical abilities are not fully appreciated within the physicist community of practice. All in all, the excerpt below is very much an expression of a strong focus on practical abilities, but it is also noticeable how Kalle talks about doing physics because it is 'fun', in contrast to, for example, a student who studies physics because science is perceived as benefiting society.

I: But why did you choose to study physics then?

Kalle: Always thought that physics is fun. But for me it has always been the experimental part, it's never been to become a theoretician or something like that...

I: What do you see as so appealing with the experimental then?

Kalle: Erm... It's this that... you can come up with solutions yourself then, and then you get...to manufacture these ideas then, even though it's not me who gets to do it, but it's the people in the workshop... But it is precisely that that's so appealing, that it's so close to working in a workshop really...

I: But you're not interested in doing the practical work yourself? You want to come up with the solutions and then give it away to someone else?

Kalle: Well, I'd like to do it all myself, but I'm not allowed to so to speak. It's the workshop that does the manufacturing, then, and that's a pity. 'Cause I can do it just as well as they can.

Throughout the interview Kalle returns to the importance of practical abilities, both in terms of how he does physics and how he thinks that physics ought to be done. For the most part Kalle's positioning of himself as a physicist takes place through a positive association with the practical aspects of the doing of physics; practical skills are something he values and takes pride in having. However, there are also occasions when Kalle is positioning himself in opposition to an approach focused on looking for 'too much' understanding (in contrast to quickly finishing the laboratory exercise). For example, Kalle related how he found working together with a particular student in the laboratory quite annoying. Since Kalle had just previously told me about how important it is to be prepared before you come to a laboratory session, I drew the conclusion that what annoyed him about his fellow student was that this person was unprepared:

I:	Aha, someone who was unprepared?
Kalle:	No, but was looking for too much understanding all the time.
I:	Ok.
Kalle:	Then it just gets frustrating in the end.

As Kalle's answer shows my conclusion, however, was incorrect. Rather, what annoyed Kalle was that the student he was working with was looking for too much understanding, was focusing too much on the analytical aspects of the laboratory exercise. Thus Kalle is positioning himself as a physics student both through an association with a practical approach to laboratory work and to some extent also through a distancing from a 'too' analytical approach. However, Kalle's participation in the physicist community of practice is more complex than that; despite his strong focus on the practical aspects of doing physics there is a tension in his story between the importance of the practical and the analytical aspects of doing laboratory work.

As mentioned before, Kalle talks about how he experiences that practical abilities are not rightly appreciated within the physicist community of practice. Furthermore, despite Kalle's strong focus on practical work, there is also a certain ambivalence in what counts as physics; he does value his practical skills and his independency, but does not see these skills as belonging to the practice of physics *per se*. When we discuss whether it would be beneficial to have more open exercises in the student laboratory in contrast to today's often very structured ones, he says:

Kalle:	Yes, then you would learn so much more. Not about physics as such, but about working independently, solving problems on your own.

The tension between the different ways of doing physics and their value within the physicist community of practice is also present in Kalle's descriptions about the purpose of laboratory work. When asked directly what he sees as the purpose of laboratory work, his answer does not so much reflect his own thoughts on the issue, but rather what he thinks is expected by the physics educators:

I:	But in the student laboratory then, what do you see as the purpose of having physics labs?

Kalle: The purpose of physics labs I guess is ... what I imagine to be the purpose with a laboratory exercise is, everyone might not understand if the teacher's or the lecturer's or whatever we should call it, stand in front of the class and present theory after theory, and then the point is to have some labs so that you get an understanding for it, how it works, sort of as a complement to the theory.

Thus, Kalle sees laboratory work as being included in the physics education as a complement to other teaching forms, in particular for students with other 'learning styles', rather than having a value for its own sake. This is further emphasized as Kalle says that if the textbook is good enough there is no need for lectures or laboratory work – you can just read the textbook and gain the understanding of the physics from there. However, when I later in the interview return to the same issue by asking him what kind of understanding he is looking for in the student laboratory, and he answers by giving an example, the example is centred around the practical experiences, not so much the understanding of theory:

I: What kind of understanding are you looking for then, in the laboratory?

Kalle: It sort of depends on what kind of lab it is, what the purpose is. But in nuclear and particle physics, then it was, then we were mainly looking at radioactive materials, and then you should 'is this dangerous kind of', you should get some experience when it comes to radioactive materials, that it is dangerous. You looked at it 'oops, this was very radioactive and this wasn't so dangerous after all'.

I: So the most important thing with that lab was to try to work with these dangerous substances, it wasn't so much about learning about decays or...

Kalle: That you went through before so to speak, beta-decay and alpha and all of that, so that was fine, so it was more about trying it out.

In summary, Kalle's narrative reflects and reproduces a basic tension in the physicist community of practice, that between the theory and the experimental work. This tension is expressed both in terms of what Kalle views as being the purpose of laboratory work versus what he learns in the laboratory, and in terms of the qualities valued by the physicist community versus what he values and sees himself as good at. One way in which Kalle is handling this tension is by positioning himself as different from other scientists and therefore not having to conform to the norms of the community. When I ask him if he thinks that his previous experiences of industrial work affects the way he works in the laboratory he says:

Kalle: Yes, I think so, I can't explain how, but I would guess so. But ... maybe not so much in the ordinary student laboratory, but at least now that I'm doing my Master's research project I can say that I work in a different way compared to how ordinary scientists work so to speak, especially since you get to do a lot of tinkering yourself. You're not ... I don't know what it is, but it feels like it's a

> difference anyway, you need ... you can stand down there all on your own and just work, you don't really need any help.

All in all, Kalle values his previous experiences and takes pride in the skills he has developed as an industry worker, but he also recognizes how these experiences set him aside from the 'ordinary scientist' and is furthermore not interested in continuing in academia, at least not beyond doing a PhD.

7.2.2 Paul

Paul was at the time of our interview studying the third year of the undergraduate Master of Science programme. He is a mature student in his late thirties and his road to becoming a physics student was far from straight forward. In secondary school he describes himself as being extremely tired of school and barely passing the vocational stream he had chosen. After graduating from secondary school he worked in a number of different jobs. In his mid-twenties he was working with selling computers and realised that he wanted a deeper understanding of how the things he sold actually worked:

> Paul: Yes, how it really works, how Internet actually works, I realised it was really interesting, and at the same time I realised that I had a huge gap.

This desire to understand computers and networks together with a broad and long-term interest in physics, 'in everything from mechanics to space shuttles', made Paul decide to go back to school, to do the four years technical stream at 'Komvux' (a municipal adult education programme). After 'Komvux' he started the Engineering Physics programme, but later changed to the Master of Science programme. He describes himself as someone who learns by doing, to remember something he needs to do it himself, he says. This focus on the hands-on activities (rather than, for example, reading or listening) is also expressed in Paul's positioning of himself as a practical person:

> Paul: Well, but it's this that I've always been interested [in physics], then I've had some bad experiences from primary and secondary school, some bad teachers, some classes with problems and so on and then you end up on the wrong track and then it's like, if you can get a job and make a living then ... I guess it's different if you're academic and not so skilled at manual labour, but I'm skilled at manual labour. And then you do it for a couple of years and realize that it's really boring and no challenge, it's not that ... you can learn a new profession in a couple of years in most cases, it might be a bit cocky to say, but no matter if it's cleaning [inaudible] or doing carpentry work or, but if you've done it, you've done it and there's nothing new to it: 'aha, it's a new plumping system, it's copper instead of...'.

When Paul started university he was motivated by his interest in physics as well as a realisation that the four year technical stream was not enough to get a job, but he does not seem to have had a very clear idea of what he might work with after the Master of Science programme. However, he had not considered a university career as an option:

> Paul: And when I started here, 'academic career' , well I had never thought that I would become an academic and now I'm sitting here and am starting to become one, it's not too late yet, but ... I will probably become an academic and I never thought that I would become a researcher...

This distancing from academia is also expressed in the interview excerpt below, where he also talks about how his image of the researcher has changed:

> Paul: And, then I come closer to my goals, which are continuously changing when you do this education.
>
> I: How have they changed, the goals? Did you want to become a researcher already from the start?
>
> Paul: No, I... it would be fun, but that I should become a researcher, that's not something I thought. I've grown up with pals who are everything from construction workers to... yeah, everything, but there are very few people who've continued to higher education among my friends. And, researcher, yeah, it would be fun, but I've never thought of myself as smart enough for that, but since I started here I've realised that maybe it's not about being so very smart, but it's about thinking that it's fun, to be interested, and have discipline...

Thus, according to Paul he started his education with what could be described as a strong sense of not really belonging in the world of higher education; he had worked a number of different 'working-class jobs' and viewed himself as skilled in terms of manual labour. In this sense, Paul's story is quite similar to that of Kalle's, but for Paul the tension between different ways of doing physics is more pronounced.

As discussed above, Paul is positioning himself as a practical person and those skills are also very prominent in his approach to laboratory work. Thus, when Paul talks about himself as a physics student he focuses on his practical skills; being skilled at setting up the experiment and getting the experiments going. He also recognizes how he is weaker in the analytical aspect of the laboratory work and talks about how this is something he is working on improving:

> Paul: What I on the other hand can think ... sometimes I can think that my weakness is this thing that I sometimes have difficulties to connect measurement results and get something out of it.

I: The analysis..?

Paul: There I can think that I'm a bit weak sometimes. But maybe that's because I do the labs together with people like Emma and Tomas, who are ... who are really skilled at that, or Tomas anyway is extremely skilled, he is so bright when it comes to analysis. But when I do lab work with others I don't see it as my weakness, so it's…

I: So it depends on...

Paul: Yes, it depends somewhat on who I'm working with, if I'm working on my own or if I kind of... I don't think that I have any particular... Well, I'm probably better at connecting stuff and such, at the same time as I probably was better at it in the beginning, I think that I'm better at analysis today, after spending almost three years here... It's a smaller difference between connecting stuff, but I'm still somewhat better at connecting, I'm working on that I next year will be just as skilled at connecting as analysing...

That Paul is working on improving his analytical skills is a sign of him recognising those skills to be important in the physicist community of practice. But he also expresses how the practical and analytical aspects of the doing of physics are not necessarily equally valued within the community:

Paul: I've studied together with Jörgen too, and he's kind of such that the theoretical is better than the practical, he thinks. … and I can say that I don't think it is.

I: Like that, that kind of knowledge is higher valued?

Paul: Yes, he somehow seems to think that it's better to be a theoretician, or that it's better to... you should get away from all the manual work and I can think that, it has exactly the same value this, 'cause they are two side of the same thing kind of, it's good with theory, but if you can't tie it to the practical...

In the above interview excerpt Jörgen is brought forward by Paul as a representative of a larger group of physicist/physics students who value analytical skills higher than practical ones. For Paul these skills are complementary; as a physicist you need to be as versatile as possible, but the interview excerpt below indicates that it is the analytical skills that are more 'taken for granted' within the physicist community:

I: But if you think about the ideal students, when it comes to working in the student lab – what are they good at?

Paul: … I can't say that I think that you ought to be more practical or theoretical, but I think that you ought to be both. You ought to be interested both in the manual and the theoretical. If you're only a theoretician…

In Paul's story there is, as in Kalle's story, a tension between the different approaches to the laboratory work; between the skills and abilities he has learnt to value prior to his university studies and the skills and abilities valued within the physicist community of practice. But in contrast to how Kalle handles this tension by positioning himself as different from 'ordinary scientists', Paul tries to align himself with the norm of the physicist community, by developing analytical skills.

To this point I have mainly focused on Paul's positionings of himself, in the following I will move on to Paul's positionings of other students. Towards the end of the interview I turned to questions concerning issues of gender and physics. When Paul is asked about the woman students he describes them as better prepared and more disciplined than the men, and in general more 'duktiga'[18]:

> Paul: Yeah, at least in the beginning I think that it seems like the guys come more unprepared, the girls are most of the time much more disciplined, are more duktiga kind of, they are often that, the girls that study here. 'Now we really go on it.' Most girls who study here are quite duktiga kind of, it's not that many guys that apply… the Master of Science programme ... hasn't so high grades, but most girls who apply here have good grades.

Paul also talks about how the woman students are not satisfied with just getting an answer from the teacher but want to reason and discuss:

> Paul: 'It's like that!', 'Yes, but you can't look at it that way, no, there's no other way to look at it!' He was very stern, concise and 'bang, bang' kind of. And many of the mathematics teachers you meet in secondary school are like that too, now I think it's changing, but 'it's like that and why isn't it in another way', no they don't manage to explain that. And I think many girls turn towards that, that they don't get an explanation, 'cause they ... I think it's male and female ... Girls like to discuss things more, they want to reason more than guys.

Paul thus constitutes the differences between the man and the woman students both in terms of the woman students being more capable, and in terms of the woman students wishing for a different kind of physics, more focused on discussions.

7.2.3 David

At the time of the interview David was studying the third year of the undergraduate Master of Science programme. Our interview lasted approximately one and a half hours and it largely revolved around David's thoughts about physics education. He is very interested in educational issues, and this, he said, is also the reason he volunteered to be interviewed. He is well-formulated and keen to discuss both his own personal experiences of his

[18] 'Duktiga' literally means capable, but it carries gendered connotations that are difficult to give justice to in translation. 'Duktig flicka' would rather translate into 'good girl', a girl that in the school context is well-behaved, does her assignments neatly and correctly, and perhaps functions as a teacher's aide.

physics education as well as more general educational issues. In secondary school he took the social science stream, specialising in economy. The reason for choosing this stream was, he explains, that he was very interested in mathematics and astronomy in primary school and saw that his father in his job as an auditor worked a lot with 'numbers'. David therefore figured that a job in economics would give him a possibility to pursue his interests in mathematics. That the science stream might be the most appropriate choice for someone interested in science and mathematics did not occur to him until later, nor that astronomy actually could be a profession. After having graduated from secondary school and working with economy for a period of time, David realised that this was not what he wanted to do and so did a one year complementary education aimed at those who want to do a science degree but who have not completed the science or technical stream in secondary school ('basåret'). David describes himself as always having had a strong interest in mathematics and astronomy, but it was not until 'basåret' that he realised how tightly the two are linked. After having been repelled by the strong practical focus of primary school physics, he now met a much more mathematical physics, which appealed to him:

I: But you've always been interested in astronomy and such...

David: Yes, astronomy and maths, have always thought that it was fun with maths sort of, it's been my favourite subject in school and so on and then when I had physics at basåret, it was completely different from the physics I experienced in primary school, it was, it felt like, 'this is fun' sort of.

I: So, what was different?

David: Well, but, physics in primary school I think, it's so focused that everything ought to be practical, then it shouldn't include maths so to speak.

He is not interested in the tinkering in the laboratory; he talks about how the school teachers' efforts to make physics more hands-on rather counteracted his interest for the subject; he had wanted the physics in school to be more theoretical and mathematical. David can thus be understood as constituting his identity as a physics student both by an association with the mathematical aspects of physics and a distancing from the practical aspects; positioning himself as an analytical person. Not only does he express his own interest in physics in these terms, but this is also what he perceives to be the most common view among his course-mates, and, implicitly, the most highly valued one. According to him the focus of the Master of Science programme is the analysing:

I: So what's interesting for you is the analysing, not the doing as such?

David: No, not the doing, it's the analysing that I find interesting and that I think... somehow I've got the impression that it's like that for a lot of people at our education since that's what the focus is on and that you know when you ap-

> ply, that the focus is on the scientist who does a lot of analysis. Sure, you can end up in a lab, but still it's somehow the analysing that the focus is on, it's not the doing, it's the analysis.

David's distancing from the practical aspects of the physics laboratory becomes even more apparent as we discuss the stereotype of the 'mechanical guy', a guy who prefers tinkering to reading instructions. The 'tinkering and testing instead of reading instructions'-approach is one that according to David does not work in the student laboratory. So not only does David express that such an approach is incompatible with his way of doing physics, he positions it as incompatible with university physics.

> David: No, I don't think so, in that case I think that girls have a small advantage, I think that in general here there is no difference.
>
> I No...
>
> David: But there can always be guys who think... that... it's in a higher degree, even if it's in a smaller degree, there are guys who might think that 'hum, I don't need to read these instructions, I just tinker anyhow' like we talked about before, but that I think you discover quite quickly here, that you can't... have that attitude, but sure it might be tried a few times so to speak, that 'this I can tinker together, no problems'.
>
> I: And, that's because the guys have such experiences..?
>
> David: Have such experiences that it works, 'cause maybe it was working when I was putting together my moped when I was sixteen...
>
> I: Precisely, but here...
>
> David: Here it doesn't work and that I think you discover quite quickly...

In the above interview excerpt it is also interesting to notice how David is positioning the woman physics students, with their presumed lesser experience of tinkering, as more associated with a, according to him, more successful approach to the laboratory work. Further, David talks at length about how many people probably have an incorrect view of physics as being close to 'engineering' and 'constructing things', something he thinks may put off many women:

> David: I think it can put off many women in particular, that they think so. That you should study physics, math and such things, 'engineer' it's all about tinkering with things, kind of, and many people can be discouraged for that reason, but I don't think that there's a lot of people who come to this area because they like to tinker with things. That you realise quite quickly, that it's not what it's about. However, it can probably put people off from applying, 'cause there are so many prejudices about what it is that one does, it's not a lot of people that know what one does when one studies physics.

All in all, David's positioning of the woman physics students as more associated with an analytical approach to laboratory work is somewhat implicit, taking place for example in terms of them being attracted by a physics education focused more on reasoning and mathematics and put off by the 'engineering' image of physics.

David's devaluing of practical skills in the student laboratory is also expressed in terms of the actual 'doing' in the laboratory being something everyone could manage, what is important is to be well-prepared:

> David: I think that it is the previous knowledge when you step into the lab that is the important thing. ... But the lab as such, the experimental doings that everyone can manage, that's no problem and is there something you can't manage, then the TA:s can always help you. So that's nothing that hinders you, it's the previous knowledge that totally affects, I think, how well you know what you're doing in the lab.

But despite David's devaluing of practical skills there is, even in his descriptions, somewhat of a tension between the value of both the practical and analytical skills. While David is certain that practical skills are of no importance in the student laboratory, he does acknowledge that these skills may be important when working with apparatus in a research laboratory. Here it is also worth noticing the sharp distinction David makes between the student laboratory and the research laboratory and how he, once again, stresses the importance of reading and following instructions:

> David: No, well, if you're going to work in a lab maybe it can be an advantage, if you know how to tinker with mopeds, 'cause it's so much apparatus and then you're obviously interested in it, but to do a lab I don't think so. 'Cause it's like I said, it's more about reading the instructions than, I think you benefit more from reading the instructions than just ignoring them and trying to figure out how the apparatus works right away.

The perceived importance of practical skills in the research laboratory is furthermore a reason for David to not want to work in such an environment in the future:

> I: Could you see yourself working in a lab in the future?
>
> David: No, I don't think so, it's a bit too much, technical stuff and machines and such and so on for me to really... It's too many such things, it's a bit like, I don't mind them, 'cause they're needed and so on. I've got nothing against computers 'cause they're needed, but I'm not so very interested in it generally, it's more that I might need it...

7.2.4 Susan

At the time of our interview Susan was enrolled in the third year of the undergraduate Master of Science programme. She is a mature student in her early forties. Prior to her physics studies she worked in a number of different jobs, primarily related to health care. Ever since childhood she has been interested in science, in particular she has a long term fascination for astronomy. Her goal is to become a researcher, possibly in astronomy, but she also expresses an interest in working with solid state physics.

When Susan talks about doing laboratory work she stresses how she wants it to be well-planned with clear instructions and she values being well-prepared beforehand:

> I: What do you see yourself as skilled at when it comes to doing laboratory work? What are you skilled at in the lab? Do you like laboratory work at all?
>
> Susan: Well, most of the time, but not always, 'cause all lab instructions and such aren't, aren't I think that thought through, but if you gotten, if it's really thought through and you've had the opportunity to prepare beforehand so you know what to look for.
>
> I: How does a well thought through lab look like then? Do you think...
>
> Susan: I think that the ones we had in atomic and molecular physics, they were well thought through and the ones in nuclear and particle physics too ... that it said that 'this you need to have read', so you know, 'you know this before the lab and do these calculations and check this out a bit', so you get, prepare your assignments so you know what you're supposed to do, so you don't feel totally thrown out, so you sit there in front of a lot of equipment and just wonder what it's good for. So that ... but it's good if it's clearly stated that 'you should have read this and do some calculations on this and...' you can see step by step in the lab what you are supposed to do and that it's easy to follow the instructions on how to do things.

She also stresses the importance of mathematical skills:

> I: So there's really no particular previous experiences or knowledge that are important when one starts studying physics?
>
> Susan: You need to know maths. If you don't know maths you're going to have a really hard time.

Susan tells me how she began her university studies with a very negative attitude towards laboratory work, which she explains was to some extent due to bad experiences in secondary school and Komvux[19], where the laboratory work often consisted of demonstrations rather than student laboratory work: demonstrations that Susan also remembers as often failing. She therefore came to university physics with an interest in the theoretical rather than the experimental side of the discipline, positioning herself as an analytical person. Her lack of laboratory work experience also made her anxious in the beginning of her studies:

---

[19] Municipal adult education programme.

Susan: It was something completely new, but it was kind of fun after all, because you got used to it and realized that, no, it was not so bad and not so scary and the previous experiences ... had said.

I: Scary, in what ways?

Susan: You were afraid to ruin the equipment and that you would make some kind of mistake so that the entire lab lesson was ruined and you had to redo it and so…

Consequently, Susan, like David, constitutes her identity as a physics student both by an association with the analytical aspects of laboratory work and by a distancing from the practical aspects. In Susan's case it is noticeable how clear a distinction she makes between the practical aspects and between 'logical thinking', how they are constituted as entirely separate in a quite dichotomous fashion:

I: But you don't think it's important to be good at this tinkering?

Susan: No, I don't think so at all, but it is good with logical thinking.

Susan's distancing from a hands-on approach to laboratory work is further reinforced as she talks about her experiences from secondary school; she recalls that the boys in her class used to 'throw' themselves at the equipment whereas the girls were more reserved. However, this is an approach that for Susan does not belong in university physics:

I: The high school thing, that you throw yourself at the equipment, does that work here?

Susan: No.

I: To approach the labs in that way?

Susan: No, it's more advanced stuff, I mean, maybe it could work on the introductory mechanics course, the first lab in that course, then it works. Then you can throw yourself at the stuff and start pulling, it doesn't matter, but then…

Although Susan in the beginning of her physics studies was anxious about using the laboratory equipment, this is no longer the case. She talks about herself as comfortable in the laboratory setting, but stresses how she is not at all interested in the equipment in the laboratory for its own sake: it is what this equipment can accomplish that interests her:

Susan: No, it's not like you're afraid to touch the lab equipment. It's like at home, you have a computer... it's like the car, a computer and a car, they should work, I don't like to tinker with the stuff, I don't like fixing the car and I don't like tinkering too much with the computer, for the computers own sake, for me it's a working tool and it should work. Period.

But, despite Susan's devaluing of practical skills in the context of the student laboratory, she does express a certain interest in tinkering with her computer (even though this is dismissed as 'nothing advanced') and in digital technology. She also talks about how one as a car owner should be able to do more minor maintenance tasks, such as changing spark-plugs:

Susan: Yes, it should work, 'cause it's a working tool that should work, period. Well, I did have a period when I thought it was fun to put some more memory into the computer, but that's nothing advanced, just open the lid and stuff it in, more or less, so that... That's not much of a challenge. But I once took a course in... it's such a long time ago, I can hardly remember the name of it, digital technology I think it was. But after that, then you learnt how to do calculations on... different codes and stuff like that and then there were, we had labs where we connected AND and NAND gates to get different things to happen, that I enjoyed, then anyway, so what was on the top of my list of presents was a small, I wanted a soldering-iron, so I could sit and potter about.

I: But that's tinkering if anything?

Susan: Yeah, that's tinkering if anything, but it's not the car 'cause it's big and nasty and it should work, I could possibly... I know how to change fuses and I can fill up the liquids and so on, I don't want to do anything more on a car. OK, I think you should be able to change the spark plugs...

To some extent this can be interpreted as Susan, like David, viewing practical skills as 'uncomplicated' and therefore having little intrinsic value in, for example, the student laboratory.

7.2.5 Mia

In the interview Mia described how she, in fifth grade, saw a picture of a woman astronaut in a school book and decided that one day she wanted to become an astronaut. Later, her childhood dream transformed into a more general interest in space research. This interest eventually motivated her to apply to the undergraduate Master of Science programme and her goal today is to work within the space industry. When I interviewed Mia she was in the first year of the Master of Science programme. However, she wanted to switch programmes and had applied for several different Engineering Physics programmes, because 'as an engineer you can always get into that world [of space research] if you are capable enough'. She describes herself as not a very strong physics student, but says that her motivation to work with space research makes the struggle worth it:

Mia: Yes, it is my interest in space that has, that is the reason for it. It's not like, I don't have an aptitude for physics really, it's something I really have to work on and dig into assignments and so on, and it came with, I want to work with space and then it's that road that is the one.

When it comes to doing laboratory work, Mia describes herself as not very confident and wants it to be well-structured. It can also be noticed how she talks about the writing of the laboratory report as being important for gaining understanding:

I: So you prefer to work, what kind of instructions do you prefer, is it the more open ones or…

Mia: No! It's the more structured ones, 'do this, do this', I think that's my ... or I find it comforting to know exactly what you're supposed to do. 'Cause it's often if it's like open that you realize afterwards that 'oh', but when you're evaluating the lab and are supposed to get a certain value then just 'we didn't think about this!'. And then it's no fun.

I: Isn't that a way to learn then?

Mia: Well, it is, but on the other hand it's really frustrating and stuff that is frustrating that you easily push away sort of, and then you have to, you might not have to redo the lab, but maybe you have to go back and do something differently. I think that if the instructions are designed in such a way that you have to write a report where you gain understanding for everything even though you earlier have gotten clear instructions, that is the best for me anyway.

Further, Mia also associates herself with the mathematical aspects of physics. This does, however, not imply that she devalues practical skills, unlike David and Susan. In Mia's description of the 'ideal' laboratory student below she stresses both the importance of being analytical, structured and careful and the importance of handling of equipment:

I: If you think about the ideal student in the laboratory, what are they skilled at?

Mia: Well, they can structure their thoughts and figure out what ought to be done on a lab and really think of all the important things that need to be measured or somehow calculated and then handle the equipment, the computers and then have some grasp of the physics behind. Is thorough.

In the interview excerpt below the importance of practical skills in the student laboratory is further stressed, but a practical focus in the student laboratory is, however, not one that Mia herself can associate with:

I: Is there a difference between what's important to be skilled at in the lab compared to other elements in the physics education?

Mia: Yes, well, it's more the practical in the lab. That you don't have, those elements you don't have otherwise, it's that you have to be able to convert the theoretical into practice and tinker together all the stuff and see that it works.

I: Is that something you like? That you feel comfortable doing?

Mia: No!

I: Why not?

Mia: I'm generally bad at tinkering...

Consequently, it was Mia's laboratory partner who was left with the handling of the equipment, while she took responsibility for the computer. In her explanation of why she and her laboratory partner had this division of labour, she positions women as having less aptitude for handling of equipment in general, drawing on a discourse about gender as biological.

Mia: Aha, I had a lab partner who, I don't really know if he thought it was any fun to tinker with the equipment, but he was doing it while I was handling the computer or...

I: Why do you think you divided the work like that, that he was doing the tinkering..?

Mia: Well, that's in the genes! That's left since, I'm quite convinced that that's left since the Stone Age, that the men were out with their tools, fishing equipment and hunted and so that they developed a feeling for it that the women didn't develop, so I think it's very natural that it's still left today, that men find it easier, that their brains are more developed for such things...

Mia expresses the view that women have more aptitude than men for logical thinking and mathematics, and she consequently positions herself, as well as other women, as more associated with an analytical than a practical approach to laboratory work:

Mia: Women are more logical, women have an easier time doing logic, but that I've seen evidence of too that girls I've had in my class or so, have had an easier time doing maths than physics.

I: So what is the big difference between maths and physics then?

Mia: Well, maths is really logic, you have some kind of pattern that you work from and then it can be a number of different things that are related and then you have to pick out one thing at a time and reach a conclusion, so it's really no understanding, just routine, you have to work back and forth and have patience if it's difficult problems, whereas in physics it's much more that you have to understand relationships, yeah, if it rolls to the right and pulls down that way it will move in that direction and then you have to sort of picture things in a different way, forces and...

Consequently, Mia constitutes her identity as a physics student both by an association with mathematics, logic and analytical skills, and a distancing from having practical skills.

7.2.6 Ann

Ann is a mature student in her late thirties. At the time of our interview she was enrolled in the third year of the undergraduate Master of Science programme. In secondary school she took the technical stream and directly after graduation she started working with construction of electrical equipment. This job was extremely man-dominated and she describes it as very masculine, in terms of jargon for example, something she eventually felt did not suit her. Today her goal is to become a researcher in atomic and molecular physics. Ann was easy to interview, she seemed comfortable in the interview situation and was talkative and reflective. Being a mature student, her decision to go back to studying after having worked for several years was well thought through and something she was happy to talk about.

Despite Ann's previous experience of working with constructions of electrical equipment she is claiming to be unskilled in terms of practical work and is positioning herself as associated with the analytical aspects of laboratory work:

Ann: If you are to do labs, put some stuff in and then measure the result, that fits me really well and then analyse it, but not if you have to tinker too much yourself, try new things, and connect stuff together and such, that doesn't fit me.

Consequently, it is not 'labwork as such' that Ann focuses on, but she sees the laboratory work as a possibility for understanding of theory, something that takes place through the prolonged engagement with the physics content, the discussions about it. Ann herself summarises this by using 'studying for an exam' as a metaphor for her learning in the student laboratory:

I: What do you see as the purpose of doing labwork?

Ann: That you should understand what you're doing in a larger context. ...

I: What do you mean by understand what you're doing?

Ann: How should I put it... The labwork as such sort of... But as an example, now that we were doing nuclear and particle physics, then you've been reading about different decays and beta-decay and everything, but then it becomes so clear that, yeah, annihilation that's them and from the positron you get that... so really I see the labs as a way to study, 'cause you understand what you've earlier been struggling with, it falls into place.

I: Because you do it in yet another way?

Ann: No, maybe because you talk, you think so much about the same thing an entire day.

I: So it's not the labwork as such, it's more that you work in a group and discuss...

Ann: Yeah, more that. Not [the lab] as such, 'cause it's often very abstract, you see some curve or some line or some spectrum, most of the time I think it's... all the going through, that it's like studying for an exam kind of, 'aha, that's how it was!', that I think is almost the most important thing with the labs, that you understand the theory.

It is also noticeable how Ann in the above interview excerpt constructs understanding of theory as the highest valued outcome of doing laboratory work.

Further, Ann describes her positive qualities as a student in the laboratory in terms of being structured, both beforehand and during the laboratory work, and skilled when it comes to writing the report afterwards:

I: What do you see yourself as good at when it comes to labwork?

Ann: I'm pretty lousy I have to admit. Well, I'm positive. I inspire my friends when they kind of feel low. No, but really I'm not that good, most of the time when I do labs I'm not that good and the person I'm doing the lab with is better, but... I can be structured beforehand, that I can... I've tried to go through stuff before and then I think I'm pretty good at writing reports, that I'm good at the work afterwards. When it comes to pure labwork it's those things I'm good at.

...

Ann: I think I benefit from being quite structured, I can take... It's almost the same thing, that first I do this and then I do that. And that you've learnt along the way that you bring that to the lab room too, that you... that you do one thing at a time and that you kind of... you reconnoitre first, ok, what am I supposed to do, what apparatus should I use and how do I do that... you're really careful throughout all the steps. That you've learnt.

How Ann is positioning herself in association with the analytical aspects of laboratory work is also mirrored in her thoughts about the kind of research laboratory she could see herself working in in the future. At the time of the interview she had had the opportunity to spend some time in a research laboratory and could see her self working in such a laboratory in the future as long as the work is not too focused on 'tinkering and connecting stuff together':

I: Could you see yourself working in a lab in the future?

Ann: Yes, if it weren't too – I thought about this – if a lab is enough... ... If you are to do labs, put some stuff in and then measure the result, that fits me really well and then analyse it, but not if you have to tinker too much yourself, try new things, and connect stuff together and such, that doesn't fit me.

From the interview excerpt above it can also be noted how Ann is constituting her identity not only through an association with the analytical aspects, but also through a distancing from the practical aspects. This is further reinforced as Ann compares herself to her course-mate Mats, whom Ann takes to represent a 'practical physics student', and whom she positions as very different from herself:

Ann: Mats, he's an experimentalist! He's so much fun to do labwork with, 'cause he really gets, he might not understand the theory at all and hasn't done anything and is tired and hasn't slept and he sure starts to tinker kind of! He's so very different, he really fits in a lab!

It is important to notice that, Ann talks about a practical approach to laboratory work as 'not fitting her' and not as being at odds with physics. On the contrary, Mats with his tinkering attitude portrayed by Ann as really 'fitting in' a laboratory. Overall, Ann views the physicist community as being made up of numerous different communities, where a wide variety of people can find a home:

Ann: Yeah, that person [the very girlie woman] will perhaps not be the experimental expert, but becomes perhaps this brilliant theoretician. That there is a place for all kinds of people. That I really like, that it's sort of very open.

Despite Ann's focus on the analysis when doing laboratory work in physics, she does value practical skills in her everyday life; she thinks one should be able to do practical work and is positioning herself here as a practical person.

I: Some also talk about how boys have more experience of tinkering and stuff and how that might be important in this context...

Ann: Yes, but there I think I'm untraditional. Because on one hand I've worked in that workshop, well, I wasn't so very, I wasn't the one who got the more difficult jobs, but I worked there quite a lot, you know you stand there and connect stuff. ... So I'm pretty handy when it comes to such stuff. Have changed the break shoes on the car once, just so that I should have done it sort of. Now I've done it, will never do it again.

Further, she is very pleased that her eldest daughter is 'tomboyish', as 'traditional femininity' is for Ann seen as limiting:

Ann: I've got two daughters, both are kind of – the oldest is four years old and it's most apparent with her, she's sort of tomboyish, plays with both girls and boys and so on. And that really pleases me, you notice already now how easy it is that girls become girly and sit and draw, kind of, and have nice hair. And that's really sad. And you bring that role along all the way.

I: And that's limiting here then..?

Ann: Yes, 'cause then you've kind of, how should I explain, if you've been out in the garage and tinkered with the cars instead of sitting inside and looking pretty and drawing, than you've understood something, so when you eventually get here you dare to do more things.

As indicated in the previous interview excerpt, gender issues are central to how Ann constitutes her identity. Issues of gender were brought up by her by her right from the beginning of the interview as she talked about her previous experiences of working with electrical constructions. Here she made a clear distinction between the masculine environment in her previous workplace and what she now finds at the physics department. With reference to this earlier experience of an extremely man-dominated working environment she strongly opposes the idea that physics should be masculine. When friends visited her at the university and pointed out to her how many men there in fact were, she was surprised, since she had not experienced her studying environment as being particularly man-dominated. Nevertheless, Ann does position the workshop community and the university-based physics community as very different from each other. Not too surprising perhaps, but in contrast to Kalle's positioning of the similar communities, in section 7.2.1:

Ann: At first I was working with the electrical constructions in one place and then I moved to the next place, the next step so to speak, if you're going to go on, then I became a consultant and then I started to feel that, no wait, this doesn't fit me. Here you're supposed to be so masculine, and I can't explain, it was a totally different... jargon. And then I felt that I can't walk around and talk about ice hockey just to fit in. I was the only women all the time. In all of Sweden I think. There was one more women in Sweden who did what I did and ... yeah. Here it feels more that it's competence, not jargon, if you're good it doesn't matter if you're... how should I explain, there it was more collective behaviour that gave a go-ahead spirit and here it's more competence, you learn something and think that it's fun and get good at it and continue working on it.

I: So for you, physics isn't particularly masculine?

Ann: No, absolutely not!

I: I could imagine many women who...

Ann: No, not at all! This isn't masculine at all.

I: I could imagine many women who would say the opposite...

Ann: No, no, no.

I: ...that physics is very masculine.

Ann: No! Not! Here it's much more... no...

I: But there's after all a lot of men who study physics. It is.

Ann: But the attitude is totally different, it's much, much... when you study it feels more... how should I explain.... it's not such workshop mentality. Nothing bad about that, I loved those guys and we had a blast and it was really funny, but... no.

Eventually Ann felt that she could not fit into the 'workshop mentality' any longer, in particular since the next step in the career would put demands on her to be even tougher, she explains, and after her second parental leave she decided to pursue her dream of studying physics. However, even though she distances herself from the kind of masculinity that characterised her previous workplace, she does claim to feel very comfortable among men, often appreciating their jargon and attitude:

Ann: I like being among guys, I like that jargon and that attitude a lot of the time, I would be more nervous if it was only women. I've had a horse you know. Only women in a stable. Not so easy, I can tell you.

In the interview excerpt above it can also be noticed how Ann is distancing herself from all-woman environments and at a different point in the interview she also talks about herself as different from other women:

Ann: I can never be like normal [women] [...] So I feel very comfortable among guys...

In terms of doing laboratory work, however, Ann is positioning herself as often having taken a 'female role':

Ann: 'Cause most of the time I've been doing lab work with guys and then most often I've taken the female role, partly because I feel a bit slow.

In the interview, subsequent to this excerpt, Ann explains how this 'female role' for her is one of passivity, where she takes a step back and lets her laboratory partner take the main responsibility for the execution of the experiment.

7.2.7 Lisa

At the time of our interview, Lisa was studying the third year of the undergraduate Master of Science programme. She had her mind set on working with astronomy, preferably of the more theoretical kind. Before she started the Master of Science programme she had spent a few years working and travelling. Despite a life-long interest in astronomy, she says that it took her several years to realise that astronomy was actually something you could work with. She describes herself as a highly motivated physics student and likes doing laboratory work, seeing it as a valuable part of her education.

Throughout the interview Lisa keeps coming back to the importance of theory and analysis as she describes her practice in the student laboratory. For example, she talks about how she finds the 'follow-up' of the actual experiment, writing the report, figuring out the theoretical aspects and analysing the measurements, the most enjoyable part of laboratory work. This is also what she would recommend others to focus on, thereby stressing that it is the analytical aspects of laboratory work she considers to be most important:

I: If a friend of yours were to start studying physics, what would you tell them to focus on in the doing of laboratory work?

Lisa: I'd probably focus on the report writing, when you write down the theories behind.

Lisa's focus on the analysis is also evident in her idea of a good laboratory exercise. When asked to give an example of such she mentions one where the measurements coincided particularly well with the theory:

I: Is there a particular lab that you thought was particularly good, that has been particularly useful?

Lisa: I liked the ones in Quantum Mechanics, when it was spectra and so, transitions.

I: Why?

Lisa: It was so obvious then, the theory worked so well with the experiment, it was great!

I: So a good lab, that's when it can illustrate the theory?

Lisa: Yes, I think so. I learnt so much from them.

While Lisa's primary focus is on theoretical and analytical aspects of laboratory work, she does also see practical skills as valuable in the student laboratory, even though the issue of her lack of such skills is nothing she dwells upon:

I: Are there any experiences you lack, something you would have wanted to have with you?

Lisa: It's the technical, then, handling of the equipment, connecting stuff, that I'm not skilled at. Sometimes you are kind of out of phase too with lectures and laboratory work, you have to do the labs before you've had the theory. And I don't think that's good.

Towards the end of the interview I brought up questions about gender and physics and, among other things, asked Lisa to speculate about why she thinks that physics is so man-dominated. Her answer draws on a discourse about 'gender as socialisation' and she also implies that studying physics might not have been an option for her had she gone straight from secondary school to university (I will elaborate on this issue in section 9.2.4).

In all the interviews I asked the interviewees whether they think that there is a difference in how man and woman students work in the student laboratory. Several of the interviewees were somewhat hesitant to answer the question, but not Lisa. She positions many man physics students as more associated with the practical aspects of laboratory work, whereas the woman students are associated with thoroughness:

I: But if you think specifically about the laboratory work, do you think there's any difference in how women and men do laboratory work? In how they work?

Lisa:	My own experience is that, not all of them, but many men have an easier time with connecting stuff and that they've done such things more, tinkered with cars and...
I:	Cause they have the concrete experiences kind of?
Lisa:	Yes, exactly, but otherwise I think the women are a bit more thorough it feels like, that every step has to be correct.
I:	Why do you think it's like that?
Lisa:	Maybe because you haven't got experience of how it ought to be done, then it's important that it's correct.

In summary, Lisa's primary focus in the student laboratory is on theory and analysis, and she is positioning herself as an analytical person. However, unlike David and Susan, Lisa's valuing of analytical skills is done without a simultaneous devaluing of practical skills.

7.3 The Graduate Students

The graduate student narratives in the following sections are constructed in a similar fashion to the undergraduate student narratives.

The focus of these narratives is on the students' approaches to laboratory work, in the student as well as the research laboratory, and their positionings of themselves and others in relation to the perceived boundaries of the physicist community. As in the last section, each student narrative is started with a short biographical note, to 'thicken the description', aiming to enhance the transferability of the results.

7.3.1 Cecilia

Cecilia talks about studying physics as something that was never a difficult choice for her. One of her parents is a physics and mathematics teacher at secondary school and science was ever-present in Cecilia's childhood. Today, not only Cecilia, but all her brothers and sisters work with science or technology. For Cecilia the science stream was the obvious choice in secondary school and a fascination with relativity theory and quantum mechanics made her choose to study physics at university. At university she did the Master of Science programme, specializing in material physics. Today Cecilia is a PhD student in experimental physics, working at a physics department. At the time of our interview she had been working as a PhD student for about six months. The interview lasted for about an hour; Cecilia seemed relaxed in the interview situation and was very talkative, giving long, detailed answers to my questions.

When Cecilia described her practice in the student laboratory this was very much focused on her confident handling of the laboratory equipment. In the interview excerpt below Cecilia says that she did not prepare much for the laboratory exercises, was unafraid to fiddle around with the equipment and that it was the practical aspects of the laboratory work she found most interesting:

> I: If you think back on the student laboratory, how did you work there, what ... for example, what parts of the laboratory did you focus on and what did you see yourself as good at and so on?
>
> Cecilia: Erm ... How do you mean then with different parts of the laboratory?
>
> I: Well, the laboratory work, I mean, there are, you are supposed to prepare, you are to write a report and...
>
> Cecilia: Ok, I really sucked at preparation, it was always the last minute to read stuff through, just before ... 'Some preparatory assignments, yeah, it was, damn.' So it was always the last minute. Erm, the laboratory report I sucked at in the beginning, was good at when I finished [the education] or I didn't suck, but ... quite bad. When I look at my first labs, then just, erm, 'god how awful'. But ... just passed like that. Erm, and then I was skilled at ... well, when I'm doing laboratory work I'm good at daring to press all the buttons, kind of, if you have an oscilloscope 'I haven't used an oscilloscope in two years, turn a bit here, turn a bit here, damn, look, picture!'. And so on. And then I have a tendency to get strange results I can't explain, but ... I always thought the practical things were fun, but as soon as there's a computer it's a bit more scratch your head and think about how things work.

Cecilia started her university physics education with the intention of getting an undergraduate Master's degree and then starting to work in the industry. She had absolutely no intention of staying at the university. This, she says, was partly due to her growing up in a small town in a rural area of Sweden where few people went on to higher education. Many of her friends were working in industry and such, rather than studying, and academia consequently felt like an unknown world for Cecilia, whereas industry was familiar ground. However, during her Master's research project she realised how much she liked doing research and decided to try to get a PhD position. Today she talks about herself as an experimental physicist, with an explicit focus on the practical aspects and the connections to reality:

> I: What's so appealing with the experimental then? If you can elaborate a bit?
>
> Cecilia: Mm. What can it be, well, I think it's ... the connection to reality, or well it's in the theory too, but it has to be ... And that you get to, kind of, press some buttons so to speak and see that something happens and think about how to solve that problem. In reality. Not in the world of the formulas.

However, when asked what characteristics are good to have for a physicist, her description stresses the importance of being analytical and logical as well as the need for creativity, and how being pedagogical can be an advantage. But, she does not seem to notice that this description is quite different from how she describes her own practice in the laboratory and, without hesitation, regards herself as fulfilling the description of the physicist as analytical, logical and creative. Cecilia's willingness to pinpoint physicist characteristics further shows that she experiences herself to be in a position where she has legitimacy to define the physicist. As will be further discussed in section 8.3.1 Cecilia does also position herself as a physicist, although she is not too keen on describing herself as a typical physicist, apart from her being a bit of a nerd. In fact she is not even too keen on describing the typical physicist at all:

I: Do you see yourself as a typical physicist then?

Cecilia: Well ... What's a typical physicist ... Erm, I think that my picture of the typical physicist is changing now that I've been on conferences and courses and so on ... 'Cause before it's been more this general picture of a typical physicist as some boring human in a corduroy jacket who is sort of confused [we are both laughing] so, but now I see it more as, I don't know what you should call a typical physicist ... more than interested in physics.

I: How has the picture changed, before you saw…

Cecilia: From being generally boring to kind of fun actually. And more ... yeah, there's. I think my prejudice has changed from being some kind of prejudice about boredom, corduroy jackets, to not really having any prejudice at all. I've seen so many different pictures, that I can't really place them in a box any more. Or I've met so many physicists, that physicists stop being a box so to speak, if you understand what I mean, so that, instead you place the physicists in different sub-boxes.

I: Then it's a bit difficult to say if you yourself are typical or not…

Cecilia: Yes, precisely, precisely, absolutely ... But you have to be a bit of a nerd kind of, be able to focus on one thing and not ... and that I've got.

In the above interview excerpt we can also see how Cecilia's view of the physicist has changed as she has met more physicists, something that will be further discussed in section 9.3.1. Cecilia is also careful not to give the physicist a gender, talking about the physicist as a 'human'. Later in the interview she does however reflect upon how it may be easier for her to fit within the boundaries of physics than for her sister, as Cecilia describes herself as more 'laddish'[20] than her sister:

I: How do you experience being a woman in this environment?

Cecilia: Erm, well its ... I don't experience any problems actually. That I can think of. I know that my sister has had more problems with it, she gets really tired of being in minority all the time, so to speak, but, I don't know, I'm more laddish than she is so that…

[20] In Swedish: 'grabbig'

Thus, Cecilia is very much aware of the 'gendered boundaries' of physics, as will be also seen in section 8.4.4, where she notices how she as a woman physicist has broken the woman norm by choosing physics.

7.3.2 Karin[21]

Karin has got an undergraduate Master's degree in engineering physics. Her Master's research project, which could be described as being in experimental physics, was done while working for a company. After she finished her Master's degree she worked abroad for a couple of years, in a company similar to the one where she completed her Master's research project. This employment was, however, very uncertain and when the opportunity to apply for a PhD position at the University arose, she took it. At the time of our interview Karin had recently begun her PhD studies, at a technological department. Unlike the rest of the students interviewed, Karin is working with an interdisciplinary project and has large freedom in defining her research questions. The applicability of her research, as well as the broader perspectives opened up by the project's interdisciplinarity, are important for Karin. However, she also feels that it is stressful not to have a clearly defined project, with clearly defined outcomes in terms of publications etcetera, as is typically the case for a PhD student in physics.

When Karin talks about her practice in the student laboratory, she describes a student who is structured and does the measurements carefully. From her descriptions about her work in research laboratories it is not clear how she is working in that environment, but the primary difference, she says, between working in a student laboratory and a research laboratory is that the latter is characterised by having a larger independence:

I: What's the difference if you think about the student laboratory during your education versus the lab you're working in now or that you worked in as a Master's student?

Karin: It's less controlled, 'cause the student lab is most of the time 'do A, do B, do C' kind of, that it's really, 'cause you only have so much time and you still have to reach some conclusions in that time. So it's much more controlled, whereas when you do labs, both like I did in [country] and like I've done here, it's more like 'we try think, oops, we do this instead'. So I think it's more like there is no key, there is no scheme for how to do it, you have to find it out yourself.

As an undergraduate Karin completed the Materials Engineering program. This was a choice that she says involved a lot of hesitation, as she figured it was only for 'very smart people', and even though she had really good grades she saw herself more as ambitious than smart, and was uncertain that

---

[21] The context of the interview with Karin is described in a vignette in section 6.2.

she would cope with an engineering program. This feeling is also reflected in her description of her current PhD project, which she describes as less advanced than her PhD colleague's project. It is not, she says, as technical and specialised as their projects. Overall, Karin is not positioning herself as a physicist, or as any kind of profession for that matter. According to Karin you need to be much more specialized than she is to position yourself as a physicist:

I: If you were to describe then, to someone who's not in this world we're in, how you learn your profession, or how you have learnt your profession, how would you describe it?

Karin: Erm [Karin's laughing] I don't know if I know a profession, it feels like I'm still learning to learn, or maybe, I don't think, well it is a job, I do go here and it is a job sort of, but it's not that I have to do A, B and C in that order, but...

...

I: How do you view your professional role today? How would you define yourself?

Karin: Erm [Karin's laughing] you mean, would I call myself a student or?

I: Would you call yourself an engineer, would you call yourself a physicist, would you call yourself a PhD stu...

Karin: PhD student I would call myself.

I: Do you see yourself as a physicist?

Karin: No, I don't, I think.

I: Why not?

Karin: Erm, 'cause I don't think that I've studied, well I've studied physics, but not as a physicist, feels narrower. I've studied so many other things too, so I don't think I've specialized enough in physics.

Thus, for Karin the position of physicist is a very narrow one, and the boundaries of physics are tightly set. Her first thought about who she would define as a physicist is an older, man professor at her department:

I: Who would you define as a physicist?

Karin: Erm, spontaneously ... then ... a professor at our department who is just about to retire. He feels like a physicist, he's of the old school, done physics and then graduate school and so on. It's like, the spontaneous thought of a physicist and I don't know if it's ... that it feels more like older men. Which sounds really sad that you feel that way, but ... 'cause others that come [to mind], like a secondary school teacher I had in physics, he was also a physicist for me, because he was so, yeah...

Thus, it is easiest to be positioned as a physicist for (older) men; in particular she associates the physicist with the stereotype of an older, man professor. What Karin describes in the above interview excerpt is most certainly a gendered boundary, something that is made particularly clear by her positioning of her man physics teacher in secondary school as a physicist. In all likelihood Karin has studied more physics and has more experience of physics research than the physics teacher, but still she is positioning him as a physicist, but not herself. Not surprisingly, the association of the physicist with an older man is not unproblematic for Karin, and she also acknowledges that it is sad that she makes this association. When asked to clarify what characteristics are important for positioning someone as a physicist, she brings up her colleague Ann:

I:	So how is the typical physicist? What characteristics and so on does it have?
Karin:	Well then I have to say that I do think that Ann for example is a physicist, she's done the Master of Science programme and she's interested in, sort of, things on a completely different level than I am, I look at the larger system perspectives whereas she's more interested in band gaps and electrons that jump, more on that level.

How Karin is positioning Ann, but not herself as a physicist, further shows how the degree is not the only thing that matters for making the physicist positioning available; Ann and Karin are in fact working at the same department and will eventually gain the same kind of PhD degrees.

7.3.3 Hanna

Hanna grew up in a small town in the north of Sweden. Neither of her parents worked with science or technology, but she describes her father as being interested in science, also another relative works as an engineer, and she credits them as sources of inspiration for choosing the science stream in secondary school and later an engineering education. When the interview took place Hanna was enrolled in the Energy System engineering program, a programme she chose because she wanted to combine studies in physics and mathematics with a broader societal perspective, but also because she perceived it as less demanding than, for example, the Engineering Physics programme:

I:	Ok, why did you choose it?
Hanna:	I first thought that it's quite, it's got both maths and physics but has quite a broad societal…
I:	Ok.
Hanna:	And some of that and some biology and some environmental aspects.
I:	Yeah.
Hanna:	That was my thought in the beginning.
I:	So what attracted you to that compared to, say, another engineering education?
Hanna:	It feels a bit, oh, 'can I cope with it'. A bit like that.

I: Ok.

Hanna: That Engineering Physics sounds so very difficult…

I: Yeah.

Hanna: A bit like that and then also ... I'm kind of interested in these environmental questions.

I: Yes.

Hanna: And then I thought, there you saw that they at least had a thought about bringing environmental issues up and discussing them a bit.

The interview with Hanna was one of the shorter interviews, lasting for about forty minutes. Especially in the beginning of the interview she seemed somewhat tense, giving relatively short answers to my questions. At the time of the interview Hanna was about to finish her Master's research project[22], at a physics department. Her project mainly involved making simulations, but she also participated in doing some experiments. It was Hanna's supervisor that informed her that I was looking for interviewees working with Master's or PhD projects in experimental physics. Hanna contacted me, but was nevertheless unsure whether her project could be described as experimental physics. At first I thought that this was because the project was more geared towards simulations than actual experiments, but it turned out that this was only part of the reason:

I: So how come you're doing a Master's research project in, I don't know if you look at it as experimental physics, but maybe one can do that, or?

Hanna: Aha, I don't know.

I: Is it that do you think?

Hanna: Well, I suppose you can say that.

I: Yeah. Was it anything in particular that attracted you to that?

Hanna: No, not really.

I: You're not doing the most experimental project of the people I've interviewed, so that question is a bit… [we are both laughing]. Do you see yourself as a physicist then?

Hanna: No, not really that either [laughter]. Actually. I've never thought about it like that.

I: Why not?

Hanna: ... Well ... It may also be the education I've chosen, that then it's not really the same as ... as, for example, Engineering Physics.

I: But the other people in your research group then, are they physicists?

Hanna: Yes…

[22] In this thesis students enrolled in the undergraduate degree programmes who are working on their Master's research projects are taken to be graduate students.

Thus, it is not only in positioning herself as an *experimental* physicist that Hanna expresses difficulty, but also in positioning of herself as a physicist in general. She rejects the physicist positioning with laughter, saying that she never thought about herself in those terms, even though she is willing to position the rest of her research group as physicists. Thus, the positioning as physicist is for Hanna a relatively narrow one, only available to those with a more specialised physics education, like Engineering Physics. Of course there is no need for the physicist positioning to be the most desirable one for someone who, like Hanna, is coming from a more interdisciplinary background. But rather than claiming an alternative positioning, as, for example, an interdisciplinary engineer, she says that no 'professional' positionings are available to a Master's student:

I:	How would you define your own professional role then?
Hanna:	... No, I don't know, really.
I:	No...
Hanna:	Actually ... no, but it feels like that when you're doing your Master's research project, you're nothing yet [laughter].
I:	But you've almost finished your project?
Hanna:	Yeah ... but it feels a bit like, you should have more experience or do more things in order to ... call yourself something, I thought.

Hanna's unwillingness to position herself as a physicist, or anything else for that matter, also manifests itself in a reluctance to talk about useful characteristics of a physicist, even though I here open up the question to include engineers and similar occupations:

I:	What is important to be skilled at then, as a physicist or engineer or whatever you're on your way towards?
Hanna:	... Well, it's kind of to be able to like be independent or ... be, critically examining ... I don't know. Erm.
I:	Are there any particular characteristics that might be good? Or particular knowledges? Or experiences? There are different ways of looking at it...
Hanna:	Mmm ... It sure is good to be like quick-witted and so and have ... I don't know ... But experiences are ... good of course. Find it difficult when one never has, has never worked it feels like. So you don't really know what it takes.
I:	Even if you might not have worked as it yourself I thought that you are in such an environment now, you work in such a group and I thought what you've seen there or...
Hanna:	Mm ... I don't know...

Thus, Hanna does not experience herself as having legitimacy to characterise the physicist, in sharp contrast to Cecilia (in section 7.3.1) who, without hesitation, not only lists a number of qualities useful for a physicist, but also regards herself as possessing those qualities. To be considered a physicist you, according to Hanna, not only need a PhD degree, but you should also have continued working with physics after gaining the degree. Consequently, she is not willing to position the PhD students in her group as physicists. In other words, Hanna experiences that the boundaries of physics are tightly set and a position as physicist is difficult to gain access to. This unavailability is further illustrated in the interview excerpt below, where Hanna describes how she, in the beginning of her Master's research project, felt intimidated by all 'the smart people' doing research:

I: So how did you experience starting your Master's research project? To start working in this research group?

Hanna: Hm… First I thought it was a bit scary actually, with everything all new and everything, but then…

I: What was it that was so scary?

Hanna: I picture people who do research as being so smart, so that people are at a different level, somehow, that's it I guess. But then you realise that, no, they are ordinary people too. And then it's been very nice at the department, a lot of PhD students, young people and so, that has made you feel at home.

In summary, Hanna perceives experience as what makes the positioning as physicist available to someone, unlike, for example, Karin in section 7.3.2, Hanna makes no references to gender when the physicist is discussed; she claims to not have reflected at all about her being a woman in physics. It should, however, be noted that Hanna makes a reference to feeling 'young and inexperienced' when I ask her what it is like to be a woman in a physics environment, signalling that being a woman in the physics environment could present similar challenges to being 'young and inexperienced':

I: How do you experience being a woman in the physics environment? Is it something you've reflected upon?

Hanna: … Not so much really, now that I think about it, 'cause it's more been that I've thought that I'm young and inexperienced, it hasn't been about me being… a woman. And I also think that when I look at people I have more divided them like that instead, not according to gender but more according to experience. So that actually feels good now that I think about it. That, no, I haven't felt that anyone else has reflected about it either.

When Hanna talks about herself as an undergraduate student in the laboratory she focuses particularly on being thorough. As with several of the other students interviewed Hanna brings up independence as the main characteristic that differentiates between being an undergraduate student and doing a Master's research project, pointing out how she possibly would have wanted more preparation undergraduate education for this new independent way of doing physics in her:

I: Do you feel that your education has prepared you well enough for your Master's research project?

Hanna: ... Could imagine that you should do something similar, a bit more ... but, I don't know, maybe it's enough that you get it now, but still a bit more that you should think for yourself, or more critical sort of.

I: Is there something you feel you've missed during the education that would have helped you now?

Hanna: ... Yes, more independent work. I think there's been a lot of group-work and...

How Hanna perceives Master's students as being expected to be more independent also manifests in her not wanting to ask for help:

I: When do you ask for help then?

Hanna: ... Hm... Sometimes it's been so that you almost don't want to ask for help because you think, 'well, I should be able to figure this out myself', so that... That you might think a bit more before you ask for help.

7.3.4 Klara

Klara is a Master's student, working in the same group as Hanna, with a Master's research project that involves both simulations and experiments. She is enrolled in the Engineering Physics programme, a programme she was attracted to due to an interest in space physics. During her education her interests have changed and at the time of the interview she said that she would like to work with medical applications of radiation. Klara grew up in a medium-sized Swedish town; one of her parents is a science teacher and she has siblings who have studied engineering too. Klara was the first graduate student I interviewed; I, as the interviewer, being new to the interview protocol meant that this interview did not run as smoothly as the interviews that followed. Klara seemed relatively comfortable in the interview situation, but was not among the more talkative interviewees.

Klara expresses that she is comfortable in positioning herself as a physicist and says that this positioning was made available to her due to her working in an environment where 'everyone is a physicist'. This can be contrasted to Hanna (in section 7.3.3) who does not position herself, nor the PhD students in the group, as physicists.

During her undergraduate education, prior to her Master's research project, Klara did not feel confident or comfortable in the laboratory. Talking about her practice in the student laboratory she says that she was more theoretically oriented and focused more on preparatory assignments and report writing than the actual laboratory work.

I: How do you look at the difference between the student laboratory and doing a Master's research project? What are the differences and similarities?

Klara: Mmm… I never really liked doing laboratory work, so to speak, 'cause… things go wrong and… yeah, it's not really fun, 'cause you don't understand everything. And it's difficult and… But, now that I've been doing my Master's research project, I've been working with a set-up that I've learnt and I know where all the cords go and I know how all the knobs should be. And it's been nice that, and it's probably time that's the big difference, you have time to really get to know your own set-up and understand how everything works, whereas it's not like that in the student laboratory, 'cause you have two hours or whatever it may be. And then you don't really have the time and you also get a bit 'asch' you can leave off all the details and try to understand some kind of wholeness instead.

I: So what have you brought with you from the student laboratory to the Master's research project?

Klara: Some experience I think. You get used to the oscilloscope and get to try..

I: What was it you didn't like about the student laboratory then?

Klara: Erm, mainly I think… that it so often goes wrong, 'cause the labs are made quite clear and you know roughly what you're supposed to get and when you don't get it or then it gets… and that's not so much fun. I've been quite theoretically oriented before this Master's research project.

I: When you think back on the student laboratory, what did you see yourself as skilled at in terms of working there?

Klara: …

I: Or less skilled at?

Klara: Then, yeah… I didn't see myself as skilled at doing the labs, but rather at doing preparatory assignments or writing the report. More the work before and after than the actual lab work.

Klara describes how it became very different in the research laboratory; she is more comfortable in this laboratory environment and seems to have also embraced the practical aspects of doing laboratory work:

I: And if you compare with today?

Klara: There's a difference. I've so much more self-confidence in the lab now and I've learnt a great deal about the equipment.

I: What do you think has contributed to giving you that self-confidence?

Klara: That I've had time to learn and that I've had the opportunity to try things quite a bit on my own, without anyone looking over my shoulder and… which can be quite stressful, and it's good of course too in the student laboratory when you have so limited time you need to get help. But, to try things out a bit.

To sum up, an important theme in Klara's narrative is the difference in her practice in the student laboratory compared to her practice in the research laboratory; how she has shifted from a very theoretically oriented laboratory practice to one also including the practical aspects.

7.3.5 Ann

Ann is the only student who was interviewed twice, both as an undergraduate and a graduate student. At the time of the first interview she was enrolled in the third year of the Master of Science programme, as described in section 7.2.6. When the second interview took place she had worked as a PhD student for about six months, in a technical physics department. Her Master's research project was in experimental physics and she did it in a physics department. The PhD project is also in experimental physics, and has potential environmental applications. The environmental aspect was what made Ann apply to this PhD project.

As an undergraduate student Ann distanced herself from the practical aspects of doing laboratory work, claiming that she was unskilled in terms of practical work (see section 7.2.6). Looking at the transcript from the second interview, the shift in Ann's positioning regarding the practical aspects of working in a laboratory are clearly seen.; she describes her PhD project as 'incredibly experimental' and says that it is her previous working experiences that help her to feel at home in the laboratory:

I: Would you say that you're still doing experimental physics, or?

Ann: Yeah, absolutely, it's incredibly experimental.

...

I: So why did you go into the experimental part of physics?

Ann: Cause I'm much better at it. I'm pretty... slow, honestly speaking, when it comes to studying new things. But experimentally I feel at home at once.

I: What is it that makes you feel at home there?

Ann: Yes, but this that's logical, that if you change that this happens and... and then I think, I was working before I started studying, I was working for ten years.

In the first interview Ann expressed a somewhat ambivalent attitude towards her experiences of working in a workshop; she talked at length about not being very handy (compared to the other people in the workshop), but also mentioned that the workshop had given her 'untraditional' skills and she took pride in being able to do work on her car. When she talked about the student laboratory in relation to the workshop she focused on the differences. In the second interview, as a graduate student, Ann describes an entirely different attitude towards the practical work; drawing on her experiences of working in a workshop she is now able to associate herself with the practical aspects of research laboratory work.

During the second interview, when I cited from her first interview (she firmly refused to read it herself, but agreed that I could tell her what she had said in it) Ann seemed surprised that she had ever distanced herself from the practical work in the student laboratory:

I: You talked about that you were interested in the analysis, that the fiddling wasn't so interesting, but that it was talking about physics, that's when you learnt, when you talked about the student laboratory.

Ann: Aha, what do you mean by fiddling, what is meant by that?

I: Well, you're the one who said it, I can cite you: 'Test new things and connect stuff and fixing and tinkering and such, that doesn't suit me.'

Ann: Fixing and tinkering [laughter], it's a good thing you develop.

I: You weren't that interested in the practical work, to do things yourself.

Ann: But how interesting! 'Cause it's exactly that I like now.

I: Yes, that's what I reacted to. 'Cause you talked about Mats, how he is an experimentalist and how he always tampers with things in the laboratory and how you aren't like that at all.

Ann: Yeah, but that I think...

Thinking back, Ann points to the Master's research project as where her attitude towards the practical work in the laboratory changed. Ann describes her Master's research project supervisor as very supportive and says that he thought that she was very skilled in the practical work:

I: But it has changed...

Ann: Yes, and I think I know why.' 'Cause when you're, I think that when you are many people in a group for example, and there are other people who are more forward. And then I might step back and let other people, who seem more skilled, do it. But now that you're on your own, that I do it at my pace and in my way and asks questions and honestly speaking, that I know Patrik [the supervisor for the Master's research project] said, he thought I was really skilled in the laboratory, so I do think I'm quite skilled at it, but I think that it's probably that I step back when others are more forward, maybe... And then you don't learn anything and then it becomes unknown and scary.

I: Yes, that's sort of the impression I got when we talked before. 'Cause... this I have to cite, 'cause it's so funny when you contrast it with what you say now. I asked you what you were skilled at in terms of doing laboratory work and you said 'I'm pretty lousy I have to admit'. [we're both laughing]

Ann: Yeah, but I was pretty lousy at those labs.

Like Klara (in section 7.3.4), Ann talks about the time aspect as a major difference between working in the student laboratory and the research laboratory:

I: What's the big difference then, in doing laboratory work if you think about the student laboratory versus Master's research project and doing a PhD?

Ann: Because… I think there's so much missing… partly I was living under a totally different time pressure then, so I wasn't at all as informed about the things I was working with, like I'm no, so it's probably that when you walked into a student laboratory it was kind of 'oh, here I'm supposed to do a lab about optics or waves or something', and then you have no clue what all the stuff is.

I: No, no.

Ann: And then you hadn't got the time either, no one could stand there and tell me that 'this knob you turn like this and this one works like this, this lamp's light is like and if you change this you change the aperture'. All that important information, there's no time for it. But when you do your Master's research project, then you know that you're going to learn this equipment. And you can sit there on your own and google and you can ask, and then you find your gaps yourself, like 'I miss this and then I go and ask someone'. And you haven't got that time in an ordinary lab, then you walk in, four hours and then you're supposed to be finished. And then I think I step back if someone seems more confident. I think.

I: Yes, it's not like I'm trying to get you to defend what you said then.

Ann: No, no. But it was like that.

I: Since it's that I found interesting, that, yes, there's really a difference in how you talk now and how you talked then. [we're both laughing]

Ann: Yeah, so interesting. Yeah, but honestly, those ordinary labs, it's really, I just got myself through those.

Further, Ann describes how she experiences the student laboratory and the research laboratory as two very different environments, seeing few similarities between the two. She says that she brought nothing with her from the student laboratory to the research laboratory:

I: But, have you brought anything with you from there [the student laboratory] to the Master's research project and your PhD project?

Ann: … No, nothing. Probably because… no, then I also think, because I was at least ten years older than everyone else in the class except for Mats [laughter] and everyone was really sweet in my class, it wasn't like, but they had… I didn't want to intrude, and be the pathetic auntie, so that, they were their group and they studied and often it was me, maybe there were some odd guy who came in on a single course and then I did the laboratory work with him. So maybe it was a bit more that you just got through things, it wasn't my mates that I did the laboratory work with and it was that I felt… As I said, they were incredibly sweet, I was fascinated that they were so open towards me and so on, but I didn't want to intrude too much.

In the second interview, Ann is positioning herself as a physicist, and she describes how it was getting the degree certificate that ultimately made this positioning available to her. However, her positioning as physicist is done with a certain hesitation; she is a physicist, but an 'amateur'. Like Hanna (in section 7.3.3) she associates the physicist with being 'extremely smart', but instead of experiencing this as intimidating, as Hanna does, Ann stresses that these 'extremely smart' people are only a small subset of all physicists. While distancing herself from the 'extremely smart' Ann is able to constitute an identity as one of the 'leg workers' of physics:

I: Do you see yourself as a physicist? Or how would you define yourself? Your profession.

Ann: Well, I am a physicist, but at the same time I feel that I'm one of those amateurs[23], but they are needed too. You don't need to be a Nobel laureate to be a physicist.

I: What does it mean to be an amateur then?

Ann: That you're not... perhaps the most quick-witted, genius. It's like Henrik and I talked about, my husband, and then he said, 'but if you go out in the professional life and think, the work is done by ordinary people and then there is this clique of extremely smart, and they are often just difficult to deal with'. [we're both laughing] No, but jokes aside I'm more of that mediocre person I think.

In the first interview Ann talked at length about gender and physics (see sections 7.2.6 and 9.2.3). This issue is somewhat less prominent in the second interview, despite this interview's more pronounced focus on gender issues. In the second interview I refer back to Ann's previous claim that she felt very comfortable in man-dominated environments, and she immediately responds to this claim by saying that the women that do work at the technical physics department are very 'laddish':

I: Last time you talked about this too, that you feel very comfortable among guys and in that kind of environment...

Ann: And the girls that work there are very laddish.

I: Who work where you're working now?

Ann: Yes, it's a very crude atmosphere [laughter]. And I think that's fun, but you know, in a joking fashion, in a way that you can't be towards any girl. But you can really be like, 'aha, you've got', kind of crude, hearty and... so they might not be so female maybe.

I: How are laddish girls then?

Ann: Maybe a bit more cheeky and you can, can take tough jokes. That I think is laddish, that you're not so... having more the gift of the gab simply, that I can imagine is a bit laddish. For a woman.

In the continued discussion about man and woman, Ann distances herself markedly from all-woman environments:

[23] In Swedish: 'glad amatör'

I: If you contrast to other environments then, that might be easier, I don't know, if you think about male and female here versus in other environments you've been in, you talk about these laddish women…

Ann: Yeah, but then I know. Really girly environments, those I can imagine, 'oh, no, never'. Hospitals. Imagine, hell yeah, to be a nurse in a ward. Really, that would never work. [we're both laughing]

Thus, like several of the other woman students interviewed, Ann describes what could be characterised as a gendered boundary of the physicist community. The above interview excerpt is further an explicit example of how Ann's positioning is taking place within a physicist context; with her description she recognizes that being a nurse would not have been an option for either one of us. The context of hospitals and nursing is thus made use of as a contrast to the physicist community of practice, thereby constructing the latter as suitable for us and the former as incompatible with who she is and also with who recognizes me to be.

7.3.6 Tor

At the time of our interview Tor had recently begun his PhD studies in experimental physics. His project is focused on magnetic properties of materials; something he says attracted him because of its complexity:

I: So, why magnetism?

Tor: It's just an extra complication of physics. [we're both laughing] It's like adding the fourth or fifth dimension to everything.

This interview excerpt could further be read as Tor constructing physics as a complex and complicated science, one that gives its practitioners status, and as a physics joke, between physicists.

Tor has got a Master's degree in experimental physics. After gaining his degree he continued working as a researcher at the same department for a few years, in a position he describes as very experimental. One of his parents has a technical occupation, something that 'of course' inspired him. The interview with Tor was the only interview conducted in English, both his and my second language. This made the interview situation somewhat less relaxed, but overall Tor was quite articulate and was in particular interested in talking about his experiences of experimental work.

Overall, throughout the interview, Tor strongly identifies himself as an experimental physicist. But, despite his strong identification with the practical aspects of physics, Tor also mentions an aptitude for mathematics as being important for his choice to specialise in physics. Nevertheless, Tor's stories about doing physics are very much framed by his interest in the experiments as such; it is doing measurements that interests him, the equations in books he often skips over.

Tor expresses great pride in being an experimental physicist and says that research really cannot be more fascinating than the research he has been involved in:

I: So what's so fascinating with making experiments and being hands-on with the physics?

Tor: Well, I guess I am the right person to ask this, because I've done a lot of experiments that are really interesting.

I: Yeah.

Tor: And the scanning tunnelling microscope, I mean... the initial project I started in my Masters' thesis was to build one, so I built a scanning tunnelling microscope from scratch basically and that ended up in creating a machine that could take pictures of atoms, so you just build something, run a program and get pictures of atoms. And I mean it can't get more fascinating than that. That is just a simple fact. And I mean, a lot of experimental physics is boring, you just get some really ugly graphs, you just don't understand and stuff like that. But sometimes, sometimes you have something, something new.

In particular he prides himself in being able to handle complicated experimental set-ups. He does acknowledge the importance of 'theoretical background' for explaining the results, but in a sense he portrays the theoretical knowledge as relatively decoupled from the ability to perform the experiment:

I: Do you feel that your education has prepared you well enough for where you are today? The position you're holding today...

Tor: Yeah, I mean, yes, for example, yesterday I was doing measurements on an instrument... Had no knowledge of the theoretical background of the measurement, but still I could sit down at, I mean a really complicated experimental set-up, do the measurements, without hesitation. And get good results, and that's just because I mean, I knew basically every small part of the set-up, you have lens there, you have amplifiers there, you have some current sensor, you have a light sensor bla, bla, bla... and if you look at this you can see what you're measuring and do it... Theoretical background is something else. I mean, of course it's important if you want to explain the results, but in that case I just walked into the lab... sat down and did the measurement.

As described above, Tor confidently talks about himself as an experimentalist and it should come as no surprise that he, without hesitation, is positioning himself as a physicist:

I: Do you see yourself as a physicist? Is that how you would define your...

Tor: Yes. Absolutely.

I: When did you start seeing yourself like that?

Tor: As a physicist?

I: Yes.

Tor: ...

I: I mean you said you started with an interest in mathematics and chemistry and than at some point you drifted into...

Tor: Hm... Erm... I mean, as soon as I finished my BSc it was quite obvious I was a physicist. But before that it was a bit open I guess. 'Cause then...

But, Tor is not only positioning himself as a physicist, a position he perceives was made available by the gaining a BSc degree, he is also the only student interviewed who without further discussion is willing to call himself a 'typical physicist':

I: Do you see yourself as a typical physicist?

Tor: Yes.

I: So, what is a typical physicist? Or how come you make that definition?

Tor: I mean... The... The typical experimental physicist is doing experiments and explaining them and I've done a lot of that. I even have a homepage where I, where I describe simple experiments for people to do at home and stuff like that.

Tor also talks about himself as a 'nerd', even though he is ambivalent to using the word, as it so often, he says, is used in a degrading sense. He also seems to find it difficult to explain in more detail what he wants to capture by using the term 'nerd':

Tor: So I'm a bit of a nerd in that sense. And I've even done [...] Physicist nerd in some cases.

I: What do you put into the word nerd?

Tor: I mean, I actually hate the word nerd, I can only use it with physicists...

I: With fellow physicists, yeah... [laughter]

Tor: Especially at one point I was teaching... it was a group of young children... the school system had decided on plucking out the best students and giving them some... complicated classes at the university and I was teaching one of those classes and everybody was calling the students... people around me called the students just nerds, 'cause they were good students. Which is a very degrading word, 'cause these are just the over-achievers, the, either just students that work hard or have a little bit of brilliance or something... were a bit outgoing. So when you're calling them nerds it's a bit downgrading... But obviously it's difficult to find a word that... that grasps that you have some nerdy knowledge in physics and...

In the above interview excerpt it can also be noticed how Tor has accepted my positioning of myself as a physicist; the interview situation, being constructed as a conversation between physicists, makes him comfortable in using the word 'nerd'.

In Tor's descriptions there is a very explicit continuation from being a physics student to becoming a physicist; the difference between the student laboratory and the research laboratory is for him quantitative rather than qualitative. In both laboratory environments the focus for him is on the handling of equipment; the equipment you handle as a researcher may be much more complicated (and expensive) than the equipment in the student laboratory, but it is 'still basically the same instruments':

I: Well, back to the experimental work. I guess that's something you've done quite a lot during your undergraduate education as well.

Tor: Yes.

I: So I would be interested in, what do you see as the purpose of having labwork in a physics education?

Tor: ... Well, I mean, the training is... The training is really important, 'cause how, I mean... depends on whether you want to go into theoretical physics or experimental physics, but... if you want to go into experimental physics you have to know how to do experiments from the ground level, you have to learn on each instrument, whether it's just a simple voltage meter or a oscilloscope or something. So you have to learn this from step one, that's why I think first year experimental classes are really important, when you only have a simple pendulum or something, you're at least learning how to record time and distance with uncertainties and stuff like that, and then you move on into electronics and oscilloscopes and more complicated equipment. So you can't go into a physics lab without having done... you wouldn't understand it.

I: As you see it, what is the biggest difference between, well say the student lab where undergraduates are doing labwork in courses and so and working in a research lab? Or, differences and similarities?

Tor: Erm... I mean there is... There are experiments that are... It's nice that you're in experimental physics, then I can take examples... I mean you can take an X-ray machine and it's basically just an X-ray machine, so that's the same, it might be a bit more complicated, but... have better resolution or some extra, some extra... variable that you can change, but it's still basically the same instrument, but I mean if you finish your Masters and then go to CERN or something, you have a totally different instrument, that you have never had the ability to see in your life.

I: No, no.

Tor: So, that's a bit extreme, but I mean, there are sometimes huge vacuum systems, that you use for growing thin films for example, that you will never have access to in a Masters study or a BSc study... Because they cost several million Swedish kronor or something, but... everything that is connected to that system is the same thing that has been used in undergraduate studies, I mean, it's a computer to control some measurements or some device, some electronics, so I mean if you have familiarity with that it's easier to take the first step to...

Here it is noticeable how sharply Tor's description of the student and research laboratories contrasts with Ann's description of the two laboratory contexts as completely different from one another. Further, it is also obvious how Tor is positioning me, the interviewer, as a fellow physicist, both in his explicit recognition of me as an experimental physicist, but also in how he finds it meaningful to give a long, detailed explanation between differences and similarities of experimental systems.

7.4 Concluding remarks

In this chapter thirteen of the students interviewed have been individually introduced. The analysis has been kept close to the empirical material and I have been generous with excerpts from the interview transcripts. The principal analytical tool employed in this chapter has been that of positioning, but the student narratives have also been constructed in conversation with the research purpose and the theoretical staging of the study.

The students represented in this chapter were chosen because of the diverse ways in which they constitute themselves as physicist, and the physicist community. However, despite this there are also noticeable commonalities. In the next chapter these commonalities will become more emphasized as the interviewees are treated as a collective group and the level of abstraction of the analysis is raised with the aim of exploring the practice of physics as enacted in the laboratory.

CHAPTER 8

The Constitution of a Physicist Community: A Collective Perspective

8.1 Introduction

After having introduced each student individually, the gaze is now turned towards them as a collective group. In the following chapter I continue the exploration of the purpose of the thesis; *to explore how physics students' simultaneously constitute the practice of physics as enacted in student and research laboratories and their physicist identities in relation to this practice.* In order to bring out the students' collective constitution of the university-based physicist community, of which they are a part, this chapter focuses on empirical questions such as: How can the students' practice in the laboratory be characterised? Are there different qualities that are valued in a student laboratory and in a research laboratory? How do the students experience the boundaries of the physicist community of practice? How is the practice of physics related to the doing of masculinities and femininities? In what ways can the physicist community of practice be understood as gendered?

I begin the chapter with an exploration of the practice of the student laboratory, which is characterised in terms of Discourse models. The applicability of these physics student Discourse models to the research laboratory is then examined and a physicist Discourse model is constructed. Next, the boundaries of the physicist community of practice, as experienced by the graduate students, are explored. The Discourse models and the boundaries are then used to discuss the definition of the physicist community of practice, in particular whether the undergraduate students ought to be understood as being a part of this community. Finally, gender in the physicist community of practice is discussed.

8.2 Physics Student Discourse Models

When the undergraduate students talk about their practice in the student laboratory, the complexity of this activity becomes apparent; some portray it as fundamentally different from other aspects of their physics studies, others make no such distinctions, some focus on the practical, hands-on work, others on theory and analysis. In order to structure and interpret the students' complex depictions of the student laboratory and their own laboratory practice I have applied the analytical tool of Discourse models (see section 5.2). A Discourse model can be described as an explanatory framework that people apply to make sense of some aspect of the world, that attempts to capture some main elements of what it can mean to be, for example, a physics student. In doing so other elements are backgrounded and the Discourse model becomes a simplification of reality, embedding assumption both about what is 'normal' (for example, in terms of attitudes or ways of acting) and about what is viewed as inappropriate or atypical (Gee 2005).

The physics student Discourse models I have constructed from the interviews with the undergraduate physics students can be understood as stereotypical 'versions' of how to 'be' a physics student in the student laboratory, simplifying the complex practice of laboratory physics by focusing on some central aspects.

I have interpreted the undergraduate students as making use of two different Discourse models to make sense of their own, and others, participation in laboratory work. In short, the first Discourse model is that of a 'practical physics student'; someone whose main focus is the handling of equipment, and the second Discourse model is that of an 'analytical physics student', someone whose main focus is the physics reasoning. It should be noted that even though the Discourse models might resemble experimental and theoretical physics respectively, they are, in fact, concerned with experimental work. The Discourse models focus on what the students describe as 'suitable' ways of acting in the student laboratory, something that was brought forward, for example, in the students' descriptions of what they see as the purpose of laboratory work in physics; what they see themselves as skilled at when it comes to doing laboratory work and what skills they see as important to have and develop to be a physicist. In constructing the Discourse models I have also taken into account what the students described as, for example, inappropriate behaviour; for example, a mentioning of how an unstructured approach is unsuitable for doing laboratory work is translated into how it is important to be structured when doing laboratory work. Apart from the core aspects, 'equipment handling' and 'physics reasoning', each of the Discourse models are also made up of a number of other attributes, describing both what is seen as appropriate and inappropriate ways of acting in the student laboratory. When constructing the Discourse models I worked with all the interviews with the undergraduate students (in total thirteen).

These were read repeatedly, looking for what the students perceived as suitable and unsuitable ways of doing physics etcetera.

In some sense what I have chosen to label Discourse models do, in important ways, parallel 'the shared repertoire' of the physicist community of practice. However, starting from the construction of Discourse models rather than making claims about the shared repertoire has the advantage of allowing for contradictory interpretations of the practice of the student laboratory without defining whether all the aspects of the Discourse models are in fact part of the same community of practice. This is not to say that a shared repertoire cannot hold contradictory elements, but a Discourse model perspective makes no claim as to whether these contradictory elements are contained within the community of practice. Further, starting from Discourse models is also a way of bringing the tensions of the practice to the fore; by analysis of the interview transcripts from the perspective of constructing multiple, possibly contradictory, Discourse models I found it easier to identify tensions that may be contained within the community of practice, instead of interpreting these tensions as representing the distinctions between different communities of practice.

Next I will present the two physics student Discourse models.

8.2.1 'The Practical Physics Student'

The Discourse model of the practical physicist represents one 'set of norms' related to the practice of the student laboratory. In this stereotypical portrayal of the physics student the handling of the equipment is seen as the core of the laboratory practice. The equipment is found as being interesting for its own sake, and practical skills in terms of equipment handling are highly valued. Reading instruction, on the other hand, is seen as unnecessary and there definitely is such a thing as 'too much analysing'. In the following I will illustrate this Discourse model with excerpts from the interviews.

The Discourse model of the practical physics student is primarily centred around the handling of the laboratory equipment, the execution of the actual experiment. Paul, for example, talks about the experimental set-up:

> I: But in the student laboratory, what do you view yourself as good at there?
>
> Paul: Connecting stuff! ... I'm fairly good at connecting, connecting things together, setting things up, get the stuff working, start the measurements and stuff like that.

Kalle uses the metaphor of working in a workshop to explain what is so appealing to him about working in a physics laboratory. Note that when Kalle talks about coming up with solutions, I, from the wider context, interpret this to mean practical solutions, in other words, how to construct certain things.

I: What do you see as so appealing with the experimental then?

Kalle: Erm... It's this that... you can come up with solutions yourself then, and then you get... to manufacture these ideas then, even though it's not me who gets to do it, but it's the people in the workshop... But it is precisely that that's so appealing, that it's so close to working in a workshop really...

Furthermore, the handling of equipment is to be done intuitively rather than with the use of instructions. Thus, 'having a feeling' for the work, being intuitive, is highly valued. Consequently it is not seen as always necessary to read instructions, one should be able to just figure out or 'tell' what one is supposed to do. This is neatly captured in the following excerpt from the interview with Lars:

I: Why don't you do that [read the instructions first]?

Lars: No, I don't think it's necessary that you read 'put it there', that you can understand.

The focus on the handling of the equipment, rather than the analysis, also implies that there is a distinction between the skills needed in the student laboratory and skills needed in other physics learning situations. For example, Lars points out how you have to be more 'hands-on' in the student laboratory and have to get things going:

I: What is important to be skilled at in the lab? Is it different things than in the lecture hall?

Lars: Yes, it is, a lot of things that you should... You have to be very hands-on, things need to happen, it can't take too long to put the stuff together.

As discussed earlier, a Discourse model is also characterised by what is seen as not belonging, for example, in terms of what is seen as inappropriate behaviour. In the Discourse model of the practical physics student such inappropriate behaviour is captured as 'looking for too much understanding'; being too 'analytical'. For example, Kalle talks about how frustrating he found it to be to work with a course-mate who was constantly looking for 'too much understanding', which lowered the pace in which he and Kalle could proceed through the steps of the laboratory exercise.

In summary, the Discourse model of the practical physics student is characterised by a focus on the practical aspect of doing laboratory work; the handling of equipment and the execution of the actual experiment. Being handy is highly valued. This Discourse model may include aspects of all four purposes of tinkering described by Parsons (1995, see section 7.1), but with more focus on the utilitarian and technological purposes.

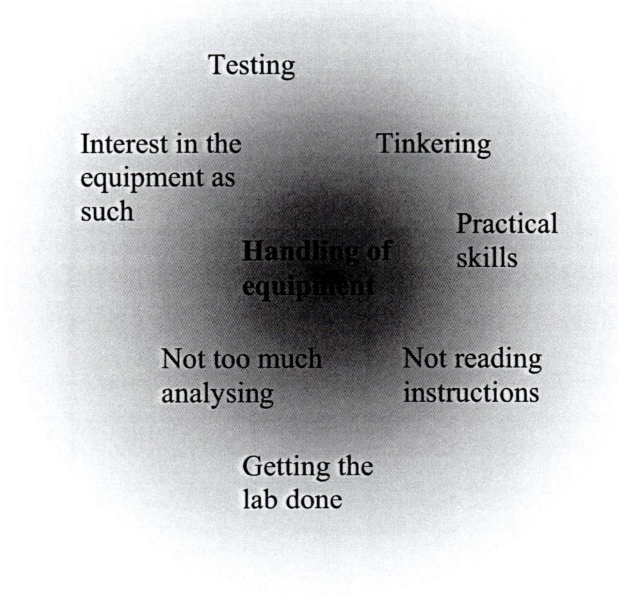

Figure 1. Schematic illustration of the Discourse model of the practical physics student. Attributes more important to the Discourse model are placed closer to the centre.

8.2.2 'The Analytical Physics Student'

In the stereotypical description of the student acting in the physics laboratory, represented by the Discourse model of the analytical physics student, the primary focus is on the physics reasoning, as taking place both through application of theory, logic and mathematics, as well as (more conceptual) discussions. The analysis is found to be the most interesting part of the laboratory exercise, rather than the actual execution of the experiment. Being skilled in equipment handling is seen as relatively unimportant, instead preparatory assignments, reading of instructions, and report writing are stressed. In the following the Discourse model is further developed with illustrations from the interview excerpts.

In the Discourse model of the analytical physics student the focus is on the 'physics reasoning', the discussion and the understanding of theory, rather than the laboratory exercise as such:

Lisa: I think it's the most fun when we go through it, when we've done the lab and are to write the report and will figure out the theory part and draw conclusions from the results we've got, such things...

The importance of the discussions is emphasised by Ann; she finds the laboratory work as such often to be very abstract, what is important for her is how she, through the prolonged engagement with the material, gains insight into the theory. Further, she uses the metaphor of studying for an exam (in sharp contrast to Kalle's metaphor of working in a workshop), a metaphor that signifies how, in the Discourse model of the analytical physics student, no sharp distinction is made between the laboratory and other aspects of the physics education. This is also brought to the fore by Susan, who also emphasizes the importance of logical thinking:

I: So what is important to be skilled at in order to be a good physicist?

Susan: It's to be able to see connections I think, to be able to think logically and see connections, to be able to connect different things.

I: Is there any difference between what is important to be good at in the laboratory compared to other stuff?

Susan: No I don't think so.

In the interview excerpts below, Dan and Lisa's descriptions illustrate how to focus here is not on the execution of the actual experiment, but rather on the preparation beforehand, the writing of the report afterwards, and the associated discussions. In Dan's description it can also be noted how he downplays the importance of the execution of the actual experiment.

I: What do you try to learn when you step into the lab?

Dan: Well, first you have to try to figure out what you're doing, but.. What you do first is to make a mathematical model, 'cause that you're supposed to do and then the thing is to get all, on one hand to get all the formulas right, yeah, and then all definitions in order to derive an expression and then check so that the expression you've derived agrees with what you measure in reality. So I think that the biggest challenge is the mathematical preparation really, if it's not a very complicated lab, 'cause I mean... Carrying out the lab, that's only mechanical stuff you're doing.

I: What to you think the purpose of labs is?

Lisa: I think it's the understanding, when you write the report you really have to go deep and read in the books, understand what it says...

In the Discourse model of the analytical physics student practical skills are consequently seen as relatively unimportant (Susan and Dan, earlier) or uninteresting:

David: It [practical physics] surely fits some, but for me it rather counteracted the interest sort of, I didn't find it fun to play with circuit cards and sit and tinker and so on, I, like, wanted to see the theory behind...

The tinkering approach to laboratory work present in the Discourse model of the practical physics student is here seen as unproductive and not suitable for the student laboratory:

David: No, it might exist, but it doesn't work here. I don't think so, 'cause it's not that kind of machines we work with, then it more becomes that you turn some knobs on random and then you might not at all get the correct values, you might not get the correct measurements, you might not turn the correct knobs at all so to speak, it's very, it's... At least I don't think you can have a feeling for something before you've seen it.

In terms of the actual practical work in the laboratory it is important to read and follow instructions, to be careful, and in general to have a structured approach to the laboratory exercise, as expressed by David and Dan:

David: I think you benefit more from reading the instructions than just ignore them and try to figure out how the apparatus works right away.

Dan: What I'm good at... Well, trying to think calmly and structured, not to stress, 'cause if you just hasten past some part of the lab, then it's easy that you miss something...

This is in agreement with what Parsons (1995) described as 'scientific tinkering'; to be more interested in understanding the theory behind the experiments than the actual execution of the experiment.

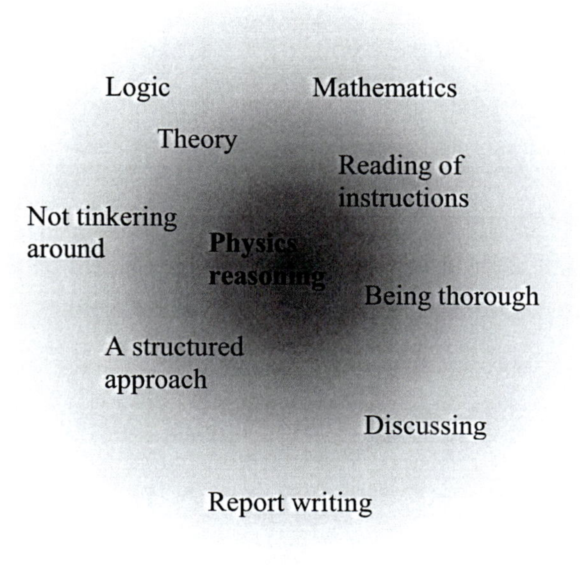

Figure 2. Schematic illustration of the Discourse model of the analytical physics student. Attributes more important to the Discourse model are placed closer to the centre.

8.3 How is a Physicist?

In the previous section I characterised the undergraduate students' approaches to laboratory work through the construction of two physics student Discourse models. In this section I use the interviews with the graduate students to explore how applicable these Discourse models are to work in a research laboratory, and what additional qualities may be needed in such an environment. This is done with the aim of constituting one or more physicist Discourse model(s).

In the interviews the graduate students' ideas about how physicists are, or should be, were primarily explored through the questions: 'How would you describe the typical physicist?' and 'What characteristics are important to have as a physicist?' However, the issue was also brought up in relation to various other questions. Here, I am not explicitly exploring characteristics tied to laboratory work, but as all the students interviewed in the second round of interviews were involved in experimental physics, much of the discussion is nonetheless tied to the laboratory environment.

8.3.1 The Physics Student Discourse Models and the Graduate Students

In general, aspects of both the Discourse model of the analytical physics student and the practical physics student were present in the interviewed graduates students descriptions about how they do research in physics.

Cecilia, for example, captures the essence of the Discourse model of the analytical physics student in her reply to the question about what characteristics are good for a physicist to have:

> Cecilia: Good to have?
>
> I: Yeah.
>
> Cecilia: Well ... analytical ... logical ... Erm, what's it called, creative. Erm ... what else can you be. It's a clear advantage if you're pedagogical, but you don't need to be. I'm thinking about my first laboratory lesson for example. ... erm ... apart from that I don't know.

Here it can also be noted how she, without hesitation, not only experiences herself to be in a position where she can actually define the physicist, but also regards herself as fulfilling those characteristics:

> I: Do you recognise yourself in those characteristics you just described?
>
> Cecilia: Yeah, sure. Or the creativity is a bit up and down I guess. I'm not creative at eight o'clock in the morning [laughter] … at midnight on the other hand.

However, when Cecilia talks about herself as a student in the student laboratory she rather talks about herself in terms of the Discourse model of the practical physics student; how she was bad at preparing before a laboratory exercise, but skilled at the actual doing in the laboratory and, for example, dared to try new things with the equipment.

Tor talks about himself very clearly in terms of the Discourse model of the practical physics student, both when it comes to the student laboratory and the research laboratory. Consequently, he identifies strongly as an experimental physicist.

> Tor: If you're going to experimental physics, if you like experimental physics or if you don't like theoretical physics. [laughter] Or a mixture of both.
>
> I: Ok.
>
> Tor: And I mean, theoretical physics… in many books and articles I read I skip over the equations, it doesn't really matter to me.
>
> I: Ok.
>
> Tor: It's more the experiment, the measurements and the stuff like that. That's, I mean, that's what I want to do.

Marina is an example of a student bringing aspects of both Discourse models to the fore. Like Tor, she identifies strongly as an experimental physicist:

> Marina: Sure, it's important to interpret the results, but ... it's still the results and observations that's the starting point.
>
> I: So why did you end up in experimental physics?
>
> Marina: Because I want to see and feel and build! It's a bit like Lego for adults, where every component costs at least a thousand Swedish crowns. Kind of needs ... precision and concentration and a great deal of accuracy. And that I find to be challenges.

The precision and accuracy mentioned by Marina may sound like aspects from the Discourse model of the analytical physics student, but I would argue that this is not necessarily the case; what she describes is a precision and accuracy in handling and even in building of the equipment. Compare Marina's utterance to the one by Karin below:

> Karin: Thorough and structured, that was like mine ... 'cause I thought it was better not the hasten of, but you do it [the lab] sort of, you take your measurements or whatever you're supposed to do, that you. Thorough and structured, to sum up.

Here the precision and the structure concerns the taking of measurements, the following of the instructions, not the explicit handling of equipment; in other words, she is describing the analytical physics students' structured and instruction-following approach to laboratory work.

However, Marina does bring other aspects of the Discourse model of the analytical physics student to the fore; both for her as an undergraduate student, and now in her Master's research project, the discussions are essential:

> I: So, have you asked a lot in the lectures and…
>
> Marina: To understand ... but the most important thing is that you really talk about it so that you understand what is going on.
>
> …
>
> I: Is that important for you now in the Master's research project too, the discussions.
>
> Marina: I think it's very important to hear others points of view and…

In summary, individual aspects of both physics student Discourse models are expressed by the graduate students interviewed. But in contrast to the undergraduate students who often discuss the two Discourse models in opposition, thereby constructing them as being mutually exclusive, the graduate students are rather describing a highly versatile physicist, skilled both in handling of equipment and physics reasoning. The notable exception is Tor, who talks about having a focus on the measurements and *not* the theoretical considerations. However, Tor also talks about how it was an aptitude for mathematics that attracted him to physics in the first place, without seeing this as conflicting with his current quite practical approach to physics.

In a sense the graduate students are capturing the 'positive' qualities from both Discourse models; they are talking much more about how to be a physicist, including both the importance of equipment handling and analytical skills, than about how *not* to be a physicist. But, apart from the attributes found in the two physics student Discourse models, the graduate students also bring two other important characteristics to the fore; independence and a certain 'nerdiness'. I will discuss the need for independence first and come back to 'the nerd' in the following section.

8.3.2 Independence – to a Certain Degree

The one thing that most of the graduate students talk about as separating their PhD or Master's research project from their earlier physics studies is an increased need for independence. Like several of the other graduate students interviewed, the PhD student Cecilia talks about her Master's research project in experimental physics as where she learnt to work more independently. In contrast to the student laboratory, with its highly structured practice, the research laboratory is perceived as offering much larger opportunities for planning one's own experiment and taking one's own responsibility:

I: What was it that happened during the Master's research project that made it possible for you to identify more as a physicist?

Cecilia: Yes, but you get ... I felt that I learnt so much during the Master's research project that I hadn't learnt during the undergraduate studies, to really feel that you can do a job and do it properly, so to speak. I do think it would be good to get more practical experiences during the education, so that we really go to experience the job so to speak. Like nurses or pre-school teachers or whatever, that you really were out in the working life. During the Master's research project you get to do that and you realise that 'yes, I can plan experiments like this and learn, to be independent and everything, has only gotten a certain introduction, so that...'.

Ann, also a PhD student, describes her entire PhD project as very much characterised by independence:

I: Does your supervisor do similar things as you do too, or? Do you sort of work on his project?

Ann: Yes, or he's incredibly ... he's mainly travelling around and lecturing and he's not at all involved in the daily work so to speak. And then he's the head of the entire division, so if you put it like this, when I got my room ... in the beginning of August, but after about a month he was checking in on me to see how things were going. And that suits me just fine, I prefer to contact him if I want, myself, and sort things out myself and go to him with questions. But I don't

	think there are so many that would like the situation, 'cause he's very, if you put it like this, I knew it was going to be independent work, but it's ... it's extremely independent in my project. But I like that.
I:	So, how's the supervision organised? Is it organised?
Ann:	Eh, it's probably so that I set up a meeting with him and then I come with my questions and we check things and then I find what to focus on and then I start doing things and then we talk about it again. But it's on my initiative, all the time.

However, the independence has its limits, it is after all her supervisor who defines her project even though he leaves the finer details to her:

I:	So he doesn't tell you that this is what you should do?
Ann:	Well, absolutely, he has given me directions that 'this you should do and this', but to figure out what it really is, that I have to do myself. And what the focus was, but ... So you kind of have to find information on your own and then sit down and talk to him and then [inaudible] some more and then you do it, and then you come back. So, that's how it is. So, as I said, it sounds negative, but it feels great to me, I would probably be more stressed if someone were on me about [what to do], so...

The Master's student Erik talks about independence as the thing that separates his current research project from working in a student laboratory. In the current project he needs to take much more responsibility than he did in the student laboratory. There is no longer a 'template' to follow, as he experienced in the student laboratory, and it is up to him to learn methods, find related research, and in general set the research up:

I:	What are the similarities between the student laboratory and what you've done now during the Master's research project and what are the differences?
Erik:	Mmm... Yeah... Some, it's not so much from the student laboratory that I directly used in the Master's research project if you... Erm... Above all, there's no template to follow now in the Master's research project. Like there was then.
I:	So how would you describe the purpose of the student laboratory versus the purpose with the Master's research project for someone who's not a physicist? What would you say?
Erik:	Yes, that you can try things more sort of and think for yourself what it is you want to get done and what you should do and think about how to do it. Of course there's, you have a supervisor who tells you what to do and so on and it's not like I figure out for myself 'aha, I should use a laser' [we're both laughing] but... You still get to do more things on your own and study the techniques and how it works and... Yeah, find it yourself in the literature and prepare everything and... You get to take more responsibility for what you're doing.

All in all, both Ann and Erik are describing their project's need for independence and a willingness to work independently, but it is important to notice that it is not an absolute independence and, further, it does not imply individuality. Ann and Erik, as well as all the other students interviewed, with the exception of Karin, work with well-defined research projects; they take part in larger research collaborations where their particular projects make up a well-defined part. As is customary with PhD and Master's research projects in physics, the research area, as well as the research methods, have most often already been well established. Independence consequently means being able to work on your own and take the initiative, but within fairly well-defined limits. As expressed by Erik, 'it's not like you figure out yourself that you're going to use a laser'. This is said with laughter, demonstrating the absurdity of making your own decisions about the research method. A gender dimension to this can also be added: in her comparative study of undergraduate English and physics students in England, Thomas (1990) also found how individuality was not highly valued by teachers or students in physics. A consequence of this was that the women studying physics experienced a need to adapt to the man majority, felt they had to be 'as good as the men' – being different was not on option. As a contrast the minority of men studying English was able to make use of the discipline's valuing of individuality; constituting powerful identities where their minority-status was an asset, providing them with special and more interesting viewpoints.

8.3.3 'The Nerd'

Apart from the need for independence, one other physicist characteristic is frequently brought up by the students; that you need to be a bit of a 'nerd'. Like most of the students I interviewed, Cecilia found it difficult to give a clear-cut description of the typical physicist. The one thing she does bring to the fore though is that, as a physicist, one has to be a bit of a 'nerd', which to her means that one needs to be able to focus solely on one thing:

I: Then it's a bit difficult to say if you yourself are typical or not...

Cecilia: Yes, precisely, precisely, absolutely ... But you have to be a bit of a nerd kind of, be able to focus on one thing and not ... and that I've got.

The idea of the physicist as a 'nerd', often meaning an ability to focus whole-heartedly on one thing, is also brought to the fore by several of the other students interviewed. For example, both Ann and Marina talk about a need to have a passionate commitment to whatever you are working with, and not to give up. For Marina, this defines the nerd:

Marina: Well... I think you have to be a bit of a nerd to be a physicist, you really have to be quite idealistic and be able to focus on something even though it looks as if nothing is coming out of it. You sort of ... a physicist is not someone who, looks one time at something and then throws in the towel, a physicist does the experiment at least three more times to be really sure that it doesn't work. And in that respect I would say that I'm a typical physicist.

However, the association of the physicist with a 'nerd' is not unproblematic for the students interviewed. For Marina there is a tension between an absolute commitment and seeing physics just as 'an ordinary job'. On one hand she talks about the devoted 'nerd' and how she does not want to have a 'nine-to-five' kind of job. On the other hand she says that her role model is a professor who is very adamant that being a physics researcher is, and should be, 'an ordinary job'.

For Tor the word 'nerd' is problematic for other reasons; he feels it is often used in a degrading way and says that he is only comfortable in using it among fellow physicists.

8.3.4 A Physicist Discourse Model

Since the physics student Discourse models are constituted from interviews with undergraduate physics students talking about laboratory work, they are *physics student* Discourse models, and I cannot really make a claim to whether they are also applicable to professional *physicists*. However, by employing the Discourse models as tools of inquiry in my analysis of the interviews with the graduate students, I could conclude that most of the single attributes of the Discourse models were also present in the graduate students' talk about their practice in the research laboratory. Nevertheless, whereas the physics student Discourse models are constructed in a more of less dichotomous fashion, with different mutually exclusive attributes, the graduate students talk about the different attributes not as each others' opposites, but rather as complementing each other. Their talk about the physicist is much more framed in terms of how to be, rather than how not to be and they bring positive attributes of both physics student Discourse models to the fore. The physicist is, in their descriptions, constructed as being highly versatile; as a physicist you ought to be skilled both in the execution of the experiments and in the analysis of them. The graduate students thus constructed the analytical and practical aspects of doing physics as complementing rather than opposing each other. This is in accordance with how Traweek (1988), in her anthropological study of high energy physicists, found that they valued versatility highly; it was not good to be identified as either as too much of a 'desk' physicist or too much of a 'floor' physicist.

In summary, a physicist Discourse model can be constructed by bringing together the positive qualities of both of the physics student Discourse models; as an experimental physicist you should combine practical skills, and an interest in the equipment as such, with analytical skills. But, apart from the attributes present in the physics student Discourse models, two other characteristics are also frequently mentioned by the graduate students; independence and nerdiness. I have therefore constructed a physicist Discourse model that is characterised by versatility in terms of being both practical and analytical as well as being independent and 'nerdy'.

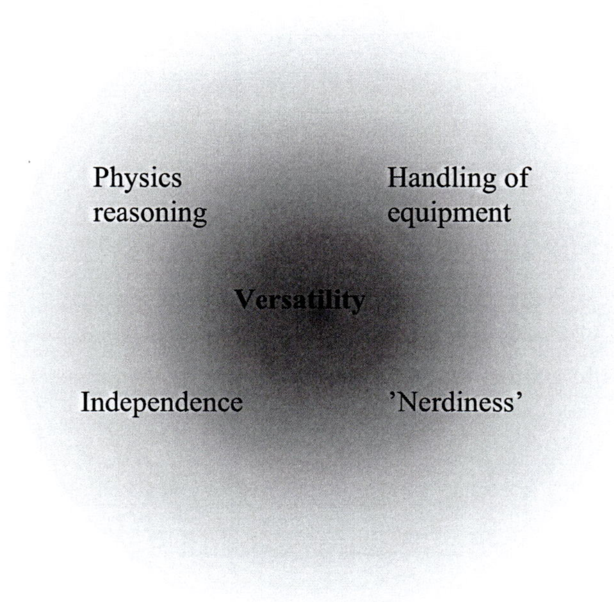

Figure 3. Schematic illustration of the physicist Discourse model.

8.4 Who is a Physicist?

In the previous section I focused on the 'inside' of the physicist community, what characterises a physicist and how the students negotiate their own 'doing physicist'. This was something I explored both in the interviews with the undergraduate and the graduate students. But, a community is not only defined by what is 'inside, it is also defined by what is positioned as being 'outside' – how the 'boundaries' of the community are set. I touched upon this issue in the interviews with the undergraduate students, in terms of what they saw as inappropriate practice in the student laboratory – and thus as practices that did not belong in the physicist community of practice. However, in the interviews with the graduate students I decided to do a more thorough exploration of how the students experience the boundaries of the physicist community of practice, in particular I was interested in the transition from student to researcher.

Consequently, in the following section the focus is on how the graduate students perceive the boundaries of the physicist community, and how they are positioning themselves and others in relation to these boundaries. In particular I will focus on how graduate students view the boundary between being a student and being a researcher and how they learn to become physicists in the transition between these two communities. Here it is important to notice that the graduate students' negotiations of the requirements for being positioned as a physicist take place in an interview context were I explicitly positioned myself, the interviewer, as a PhD student in physics and stressed my own background in experimental physics. Consequently, the interview situation is in itself constructed as physicist community, and I am able to explore how the students are constituting identities as physicists in relation to such a community, rather than in relation to the general public or to their family and friends. This also allows for a more nuanced exploration of how the students negotiate the boundaries of physics as my physicist position makes it appropriate for them to share more detailed explanations.

8.4.1 The Degree as the Dividing Line

Ann and Tor are both PhD students in experimental physics and in the interview situation they were both positioning themselves as physicists. This is a positioning they both experience as being made available to them by their completing undergraduate degrees in physics. Ann expresses this as follows:

I:	When did you start seeing yourself as, yeah, some kind of physicist anyway?
Ann:	Well, I do and I'm proud to be one...
I:	When did you become one?
Ann:	When I became one? It was sort of when you got the certificate in the mail, that 'yeah, I did actually manage this education'.
I:	When you got your degree certificate kind of?
Ann:	That, 'yeah, hallo, I got through an education and you can see that I actually passed, well, then I'm actually a physicist'. Like that you can think about it... I think.

Thus, obtaining the degree might seem to be the obvious dividing line between the student and the professional physicist; once you have gotten your degree you have it on paper that you are a physicist and can confidently position yourself as such. However, some of the students interviewed, like Erik in the next section, were positioning themselves as physicists long before they had finished their education. While Marina also acknowledges the possible importance of the degree in order to be positioned as a physicist by others, in the interview excerpt below she is still positioning herself as a physicist despite the fact that she has not finished her degree:

> I: Do you look at yourself as a physicist then? Today.
>
> Marina: Yes, even though I haven't got my degree yet, after some ten years with physics I would dare to say that I'm a physicist.
>
> I: When did you start seeing yourself as a physicist then?
>
> Marina: When I started… doing research and working independently.

For Marina it is her prolonged engagement with physics, and in particular her experiences of doing research independently (an important aspect of the physicist Discourse model), that positions her as a physicist. Thus, Marina is able to position herself as a physicist without an undergraduate degree, while for Hanna even a graduate degree is not enough to position someone as a physicist. Throughout the interview Hanna talked a lot about the importance of experience and for someone to be positioned as a physicist, according to her standards, a PhD and some additional working experience is required:

> I: You talk a great deal about inexperienced versus experienced. When is someone, according to you, experienced enough to be a physicist then for example?
>
> Hanna: Yeah… Well… That's a good question… I don't really know that either… Am thinking about the PhD students, yeah, then I think, well, something more is needed. So that…
>
> I: The PhD students in the group, do you experience that you are sort of on the some level or..?
>
> Hanna: Well, no, they are a bit higher I would say, but not so much. But, maybe if you've done a PhD and… then… and continued working, then…

In summary, an undergraduate degree (or even a prolonged engagement without a completed degree) may make the physicist position available to a student, but it is not self-evident that a student with a physics degree does position themselves as a physicist.

8.4.2 'Well, I'm Not a Chemist'

At the time of our interview Erik was studying the final year of the Engineering Physics programme. He was about to finish his Master's research project, which was carried out in an interdisciplinary research environment. Like Ann and Tor, Erik confidently positioned himself as a physicist. In fact, he talked about himself as a physicist even before I brought up the issue, in relation to my question about how he experienced his Master's research project:

Erik: Well, then I checked what they would be doing and then I saw 'aha, artificial photosynthesis, hm, what's that?'. Photosynthesis I knew about, but I hadn't been in contact with any artificial photosynthesis before, checked it out a bit to see what it was and it did sound interesting. And then I thought, 'I wonder if a physicist would be able to do something there. And then I checked it out and it seemed like there were some exciting things and then I called, no I emailed first, the head of department or what's it called, or the professor. Yeah, and contacted about Master's research project and then he said, and then he emailed back and, yes, it seemed interesting and so on. That's how it happened. So I finally started my Master's research project.

In contrast to all the other students interviewed, Erik's positioning of himself as a physicist takes place in relation to a different discipline, that of chemistry, and not in relation to being a student:

I: You talked about yourself as a physicist. Do you see yourself as a physicist?

Erik: Yes, I am. They are chemists, the ones who work there, but the group is chemical physics. So it's a bit of a mix, what's physics and what's chemistry, it's not so easy to know.

I: I've also been in that boundary area and worked, I've worked in a physical-chemistry division, as a physicist.

Erik: Yes, they call it, the department is really, it used to be the department for physical chemistry, but now it's the department for [...] and they call themselves the division for chemical physics instead. So yes, I do feel more like a physicist than a chemist.

Thus, for Erik, positioning himself as a physicist is in a sense a given, as he experiences 'chemist' as being the only alternative; that he would not position himself as a 'scientist' of some kind does not even seem like an option.

8.4.3 Physics Student or Physicist?

So far I have explored two different boundaries that the graduate students' made use of to position themselves as physicists; the one between having a degree or not, and a disciplinary boundary towards chemistry. Next I am going to deepen the exploration of how the graduate students are positioning themselves and others as physicists, or not, by focusing on the process of changing one's positioning from physics student to physicist. For many of the graduate students interviewed this was something that took place over the course of the Master's research project (as will be discussed below), but for one of the students, Erik, the positioning of himself as a physicist was available much earlier than that. As seen above, Erik started to talk about himself as a physicist before I prompted him to react to this positioning. As a follow-up to the conversation above I asked Erik when he started to see himself as a physicist:

I: When did you start seeing yourself as a physicist then?

Erik: Yeah...

I: I was thinking, in the education...

Erik: Hm... Maybe not right away anyway ... but ... 'cause ... in the beginning it's difficult to get a grasp on what you're doing, you're studying a lot of maths and many different theoretical, sort of, you don't really feel that you are doing anything. Ok, you might be doing some mechanics lab and so on, but it's still not... It feels a bit like you're following a protocol and then, check, check, yes, that's ok and that was right...

I: So when did you start doing physics then? In such a way that you could identify as a physicist.

Erik: Yeah... Started... It could be in second year kind of, then you could feel. Well, it's with the labs that you sort of feel that you're a physicist I think.

I: Ok, what is it with the labs that make you feel like a physicist?

Erik: Well... Hm... You get to do something, you get to use what you've learnt, a bit anyway, even though a lot of it is about following protocols. You still get to see a bit somehow... Some small application of it.

Thus, for Erik the physicist position was available early on in the education; it even seems difficult for him to imagine that he would not have positioned himself as a physicist during some point of his education. It can further be noticed that it was the laboratory work that made the position of physicist available to Erik; it is the doing, the application of knowledge that defines being a physicist for him. Cecilia also describes a similar experience tied to the doing of physics in the laboratory – with the difference that it was her Master's research project in experimental physics that made it possible to position herself as a physicist:

I: Do you see yourself as a physicist today, or how ... what you describe?

Cecilia: Yes, physicist. Yes, precisely.

I: When did you start seeing yourself as that do you think?

Cecilia: Erm. Right when I got the Master's research project, or, during the Master's research project. Before I was more of a physics student, so to speak, but during the Master's research project you're transformed from a physics student to a physicist kind of, when you really get the paper in your hand then, 'damn, I'm a physicist!'.

I: What happened during the Master's research project then that made it possible for you to define yourself more as a physicist?

Cecilia: Yes, but you get ... I felt that I learnt so much during the Master's research project that I hadn't learnt during the undergraduate studies, to really feel that you can do a job and do it properly, so to speak. I do think it would be good to get more practical experiences during the education, so that we really get to experience the job so to speak. Like nurses or pre-school teachers or whatever, that you really were out in the working life. During the Master's research project you get to do that and you realise that 'yes, I can plan experiments

> like this and learn, to be independent and everything, has only gotten a certain introduction, so that...'.

Thus, both for Erik and Cecilia it is the doing of experiments that made the physicist position available to them. But, while these experiments in Erik's case took place in the student laboratory during his second year of the undergraduate education, in Cecilia's case they took place during her Master's research project. And, as for Tor and Ann, Cecilia too experiences the degree as the final boundary line between physics student and physicist.

Klara, below, also describes her Master's research project as having made it possible to position herself as a physicist, but from a slightly different perspective than that of Cecilia. That Klara is almost finished with her education does play a role in gaining availability to the physicist position, but more importantly it is working in a research environment where 'everyone' is a physicist:

I:	Do you see yourself as a physicist?
Klara:	I think I've started to do it, now.
I:	When did you start doing it?
Klara:	Well, yes. After a while when I had started the Master's research project
I:	Ok.
Klara:	You are ... We have been a group where all are physicists and that ... We are physicists. So then I started, I'm probably a physicist too. [we're both laughing]
I:	What is it that has made you start to take up that identity, so to speak?
Klara:	Erm ... Yeah ...
I:	What is it that has changed?
Klara:	What has changed it that I'm more or less finished, too. That I've come closer and closer to finishing an entire education and then it has been a process, that now I'm almost an engineer, now I'm almost a physicist. And then ... Now I think I'm a physicist actually.

The perceived boundary between the physics student and the physicist could also be described in terms of the physics student and physicist Discourse models introduced earlier. By a gradually increased independence during the Master's research project (or earlier/later for that matter) the physics student is able to fulfil more and more of a physicist Discourse model and thus gain availability to a physicist position.

8.4.4 A Gendered Boundary

Last, but not least, there is yet another boundary that is voiced primarily by the woman students – that between physics and women/femininity. In this section I will stay close to the empirical material to illustrate how the 'gendered boundaries of physics' are perceived by the students interviewed, and leave the more theoretical discussion to sections 8.6.1 and 8.6.2.

Karin was unusual in the sense that when asked during the interview who she would define as a physicist, she explicitly brought up the idea of the physicist as an older man. The person she spontaneously associated with the physicist was an older, man professor at her department and she also mentioned her man secondary school teacher in physics as someone she would define as a physicist. It is also noticeable that it was not any man that Karin associated with physics, it was an older man – thereby constructing a norm for the physicist that includes both gender and age.

All the woman students I interviewed (with the possible exception of Hanna) did, in various ways, recognize how they as women had made an unusual choice in choosing to study physics. Marina expresses this in terms of how people in general think she is an atypical physicist because she is a woman:

> I: Is there something that makes you a non-typical physicist? Whatever that may be...
>
> Marina: Yeah, people think that it's unusual to be ... woman and physicist, it's still sort of 'What! You and physics!'. I don't experience it like that, but other people think I'm non-typical because I'm a woman.

Cecilia approaches the gendered boundary of physics from a different perspective. Whereas Marina talks about how you as a woman in physics 'break the norms' of physics, Cecilia talks about how you as a woman 'break the norms' of femininity by choosing physics:

> Cecilia: When you as a woman have chosen physics you've already broken the norm in some way, so to speak...

She further elaborated on this by pointing out that she, as more 'laddish' than her sister, who also is a physicist, therefore has an easier time fitting within the boundaries of physics.

Whereas most of the woman students interviewed, have, in one way or another, had to deal with the perceived discontinuities between femininity and physics (as I also discuss, for example, in section 8.6.2), the man students are not even able to talk about what it means for them to be 'a man in physics'; the overlap between being a man and being a physicist is so self-evident that it cannot even be talked about. As pointed out many times before, 'this masculinized ideology [of science] is so entrenched that it has become transparent' (Kleinman 1998). Of course not all masculinities have an unproblematic relation to the doing of physics (Paul, from the first round of interviews, is an example of this), but to possibly articulate this seems difficult.

When asked in the interview how he experiences 'being a man in the physics world' Erik reacts with laughter:

I:	So how do you experience being a man in this environment.
Erik:	Well ... there's no problem [laughter].

Tor, rather, seems to be threatened by the question; for him to reflect upon his identity as a man is so unthinkable that he at first does not even understand the question:

I:	How do you experience being a man in the physics world?
Tor:	... What do you mean?
I:	It's a question I... Is it something you've reflected upon?
Tor:	Being a man in a physics world?!
I:	Yeah
Tor:	... Well, obviously, as most physicists are men, it fits.

8.5 From Discourse Models to a Community of Practice

So far I have explored the practice of physics from two different perspectives: first by a characterisation of the 'inside' of the community by the construction of Discourse models, and second by exploring how the graduate students experience the boundaries of the community. However, this exploration also raises questions with regard to how to define the physicist community of practice; are both the physics student Discourse models part of the physicist community of practice? Do we need to understand the students and the researchers as two separate communities, or as parts of the same community? As argued earlier, an advantage with starting from Discourse models rather than from defining a 'shared repertoire', is that it allows for the inclusion of contradictory elements without having to make any claims about whether these contradictory elements belong to the same community of practice. But, having constructed two physics student Discourse models and a physicist Discourse model it is time to look at how these relate to the physicist community of practice.

I will start by exploring how the undergraduate physics student interviewed can be interpreted as relating the physics student Discourse models to the physicist community of practice. I will then move on to discuss how the undergraduate and graduate students can be interpreted as belonging to the same or different communities of practice.

At a first glance the practice of the student laboratory work might seem highly reified in terms of laboratory instructions (for an example of a laboratory instruction, see Appendix B) and laboratory equipment. At the University the physics students are often given detailed instructions and are working with more or less 'ready-made' laboratory equipment, even though some laboratory exercises are more open-ended. This 'reduction of the knowing' to reified items 'may create the illusion of a simple, direct, unproblematic relation between individual learners and the elements of a subject matter' (Wenger 1998, p. 265). But even though the reification of the practice profoundly shapes the students' experiences, the experience of meaning is, as argued by Wenger, never just a mechanical realization of a procedure. The undergraduate students interviewed do consequently also constitute quite different versions of the practice of physics as enacted in the student laboratory. For example, some of the students interviewed view reading of instructions as a central part of the practice of the student laboratory, whereas others find instructions superfluous. So despite the students interviewed belonging to quite a limited set of the physicist community of practice and describing a relatively well-defined part of the physics practice (the student laboratory), their experienced repertoire is not only heterogeneous, but also include conflicting elements. Further, in the students' learning to become physicists they develop their practices and their abilities to negotiate meaning. In other words, they 'create ways of participating in a practice in the very process of contributing to making the practice what it is' (Wenger 1998, p. 96).

An important part of the learning to become a physicist is the students' struggle to define the enterprise of physics and their reconciling of conflicting interpretations of what the enterprise is about. The presence of conflicting interpretations among the undergraduate students is perhaps seen most clearly in the interview with Paul and his talk about how his course-mate Jörgen values theoretical knowledge higher than practical knowledge, a valuing Paul does not agree with; for Paul the different kinds of knowledge are of equal importance. Even if most of the students interviewed were not as articulate as Paul about the valuing of different knowledge and skills, many of the student narratives carry a tension between the two Discourse models; do both belong in physics, or is the practice of physics confined to the Discourse model of the analytical physics student?

For some of the students interviewed the physicist community of practice is characterised by an explicit exclusion of the Discourse model of the practical physics student. This is perhaps most obvious in the case of Susan and David who both constituted the appropriate practice of the student laboratory as more or less coinciding with the Discourse model of the analytical physicist, thereby constituting rather sharp boundaries of the physicist community. However, to at least some extent, most of the undergraduate students interviewed do recognise the importance of aspects of both Discourse models, but how they position themselves in relation to the Discourse models varies markedly, showing the flexibility of the repertoire of the practice of laboratory work. This flexibility should not be understood as an 'anything-goes-attitude' with an equal valuing of the different approaches to laboratory work expressed in the Discourse models. While there is no absolute hierarchy between the Discourse models, the analytical one is most certainly the one most often mentioned as more highly valued within the physicist community of practice. This higher valuing of the Discourse model of the analytical physics student expressed by Susan and David and how Paul perceives this Discourse model to be higher valued within physics, I would argue, needs to be understood in relation to broader cultural understandings of the value of different kinds of knowledge. As discussed by Paechter (2006c), disembodied and decontextualised forms of knowledge are the ones that bring status in our society, whereas embodied knowledge (knowledge through the body) is associated more closely with subordinate groups, such as, in the study by Willis (1977), working class men. I will return to this issue when I discuss the Discourse models in relation to 'technological masculinities' (section 8.6.1.1)

In summary, there is no consensus among the undergraduate students about how the Discourse models relate to the physicist community of practice. Rather, the Discourse models are used as resources for negotiating the meaning of 'laboratory work' and constitute complex and sometimes conflicting interpretations of what the practice of laboratory work in physics entails. One way of delimitating the physics student community of practice would be to use an operational definition of the community of practice as 'what physics students do', which would then include both physics student Discourse models. Defining, or rather describing, the physics student community of practice as incorporating both Discourse models raises the question whether if this 'community' is distinct enough to be understood as a community separate from the physicist community of practice. From the interviews with the graduate students it is clear that there can be no simple and univocal answer to that question. While several of the students did talk about the Master's research project and/or the gaining of a degree as what made it possible for them to position themselves as physicists, how the students delimitate between being a physics student and a physics researcher varies markedly. There is a range from Erik who has seen himself as a physicist ever since his second year as an undergraduate, to Hanna who sees the physicist as someone with a PhD in physics and additional working experience. This difficulty to distinguish a distinct physics student community of practice is, I would argue, in fact a reason not to make such a distinction; the similarities between the undergraduate and the graduate students' descriptions of their practice are further large enough to include them both in the same physicist community of practice.

The student community and the researcher community (as experienced by the graduate students), however, do differ in at least two ways. On one level there is the relatively straightforward increase in independence and involvement in the discipline – the attributes distinguishing the physicist Discourse model from the physics student Discourse models. But on another level, there is an even more marked difference; the struggle to define the enterprise that is present amongst the undergraduate students and which characterises their talk about the physicist community of practice is more or less absent from the graduate students' talk. However, instead of understanding the physics student community and the physicist community as two separate communities, the differences can be understood by portraying the students as participants in overlapping communities of practice. Apart from participating in the physicist community of practice the students interviewed also participate in several other communities: the undergraduates in a general undergraduate student community and the graduate students in a general graduate student community (and some of them well on their way to participating in a university research community). The general graduate student community can then be said to be characterised by a larger independence and a deeper involvement in the discipline (what the interview students called nerdiness) than the general undergraduate student community. Thus, an interpretation could be that it is this transition from student to graduate student in a more general sense that is observable in the difference between the physics student Discourse models and the physicist Discourse model.

8.6 Gender in the Physicist Community of Practice

In the following I focus my exploration onto 'gender in the physicist community of practice'. First I discuss the physics student Discourse models in relation to the doing of masculinities and femininities. Next I explore how the physicist community of practice and its boundaries can be interpreted as gendered.

8.6.1 Gender and the Physics Student Discourse Models

Physics is generally seen as having strongly masculine connotations (see, for example, Traweek, 1988 and Thomas, 1990). A connection between masculinity and physics also emerged during my interviews when the students spoke about gender and physics. For example, Bo talks about how a perceived difficulty of physics can afford men status, a status women according to him are not interested in:

Bo: ...Physics was considered difficult, math... science overall is considered difficult, earlier those teachers were even considered to have a more qualified education, or how should I put it, they were considered to have higher status, back in time, and that still hangs around. Then it was of course more men

> that were attracted to... to that education then. You women haven't had the same kind of need for such external attributes.

The connection between masculinity and physics is also made by John, who when asked why he thinks men are in the majority in physics, compares his way of thinking with that of his girlfriend:

> John: My girlfriend studies the same programme as I do, well, I've compared a bit how she thinks and how I think and it's... the biggest difference is that I often think a bit differently in like kind of formulas back and forth, whereas she remembers one thing and then learns that thing and has difficulties building on it, comparing it to other things she's learnt. But she is better at learning a lot of things at the same time. I learn a few things and then I develop a theory and that is... I don't know... and to just learn a few things, it's... better if you learn languages and such, but for physics and mathematics then you should rather be able to continue building on formulas also.

However, as interesting as the association between masculinity and physics is, I would argue that one needs to go beyond the rigid dualism of masculinity versus femininity in order to more fully understand students' gendered identity-constitution. By examining the physics student Discourse models and the students' positioning in relation to them in terms of gender, I am aiming to paint a more nuanced picture of physics students' gendered identity constitution. I will do this by taking into account multiple masculinities and femininities. In the following I will discuss how the two physics student Discourse models can be seen as related to two 'technological masculinities' described by Wajcman (1991). Then, the construction of 'science student femininities' both within science education research and the interviewees' narratives will be discussed.

8.6.1.1 Masculinities in Practice

Traditionally most of the research into gender and education has focused on woman students. Over the last ten years or so there has, however, been an increased interest in issues concerning 'masculinities and schooling'. In particular this research has focused on the underachievement of working-class boys in primary schools (Skelton 1998; Weaver-Hightower 2003). The following section is also concerned with the construction of masculinities within education, but here the aim is to explore how man university physics students relate to masculinities and femininities in their constitution of physicist identities.

The physics student Discourse models do in important ways parallel what Wajcman (1991) characterised as two different technological masculinities. In her pioneer work on technology and masculinity Wajcman gives a theoretical description of technology as a masculine culture. In this description she argues that control of technology is at the core of hegemonic masculinity and she distinguishes between two different forms of technological masculinity: one based on physical strength and mechanical skills and one based on 'the professionalized, calculative rationality of the technical specialist' (p. 144). Furthermore, Wajcman argues that 'masculinity is expressed both in terms of physical strength and aggression and in terms of analytical power' (p. 145), making the masculine ideology of technology a very flexible one.

Central to both the Discourse model of the practical physics student and Wajcman's first technological masculinity is the focus on mechanical skills. The Discourse model of the analytical physics student and Wajcman's second masculinity share a focus on analytical power. Despite the many similarities between the Discourse models and the technological masculinities described by Wajcman, I do find it problematic to straightforwardly characterise the Discourse models as 'masculinities'. Firstly, because even though the Discourse model of the analytical physics student does have the same core characteristics as Wajcman's second masculinity, it is broader; the students do not solely focus on analytical power but also tie in, for example, writing skills. Secondly, I would argue that by portraying the various ways of doing physics as masculinities there is a risk of 'essentialising', that is of portraying certain characteristics as being inherently masculine.

In this context it can also be noted that Mellström (1995) has, in his research, related the two technological masculinities described by Wajcman to a Swedish context. Mellström has made an ethnographic study of two different engineering workplaces, a design and development division at a large car corporation and a small, high-tech enterprise. For the engineers at both companies, technical skills, the ability to take care of a wide range of practical problems, is seen as an important part of what it means to be a competent man. This ideal, according to Mellström, is found in a variety of social contexts, but perhaps primarily in rural areas and smaller towns and among the 'working class'. In summary, Mellström argues that, in Sweden, being practical has traditionally been valued highly and is also tightly interconnected with 'being a man' (Mellström 1995). Further, in his studies of Swedish engineers and Asian auto-mechanics Mellström (2002, 2004) shows how, across many different contexts, identification with technology is an important part of what it means to be a man. The connection between technological know-how and masculinity is also brought to the fore in science education literature. In her study of Swedish pupils in late primary school, Staberg (1994) found that the boys had a more practical approach to science, whereas the girls' approach was more theoretical. In particular, Staberg interprets the boys with parents without a higher education as seeing technical skills as an important component of masculinity.

Finally, it can be noticed that the Discourse model of the practical physics student can also be understood as related to a doing of masculinity in a different way; through its sometimes playful approach to the equipment. Mellström (1999) has characterised the world of technology as a world of 'eternal youth', where boys' childhood-play with technological toys (i.e. a certain child masculinity) continues into a professional technological world that rewards 'boyish' curiosity and inventiveness. Technology, as a men's way of life, is often founded in early childhood; the boy interested in technology is expected to grow up to become an engineer. Du's (2006) study of Danish engineering students also showed precisely this:

> It has been a 'natural' process for the male students to enter the engineering professions due to their passion for tinkering, experience of machinery, wish to work with technology and a suitable gender role to undertake engineering as a future occupation. Most of them had a clear determination to become engineers since childhood or early adolescence.
>
> (Du 2006, p. 37)

The boyish playfulness described by Mellström (1999) in the context of engineers is also often associated with physics; an example being how Richard Feynman is generally portrayed (and portrayed himself) in this way (see, for example, Feynman 1992). In her anthropological study of Danish physics students, Hasse (2002a) also found how 'playing around' (with equipment, for example) was an important aspect of many man physics students' participation in physics.

8.6.1.2 'Science Student Femininities'

In the previous section I discussed how the two physics student Discourse models can be understood to at least partly overlap with the technological masculinities described by Wajcman (1991). But what about femininities? Is it possible to perform femininity in the context of doing laboratory work in physics? Well, first it can be noted that when the undergraduate students interviewed explicitly talked about femininity, it was common for it to be in the context of how 'traditional femininity' is not resonant with doing physics. In the interview excerpts below both Ann and Lisa position themselves as being different from other women.

> Lisa: I'm not like everyone else, I walk my own road I think. I did start studying really late.
>
> ---
>
> Ann: I can never be like normal [women] [...] So I feel very comfortable among guys...

This counter-identification with traditional femininity amongst woman students in science and technology has also been well-documented in previous research (see, for example, Walker 2001, Hughes 2001 and Henwood 1998), and is further discussed in section 8.6.2. In this section I instead, focus on aspects *within* the practice of physics/science that are generally associated with femininity, drawing both on my empirical material and previous research within science education.

'She's hardworking, he's clever'

The story told by science education research about woman students' engagement in physics in general, and laboratory work in particular, could be summarised as something like:

It is important for the woman physics student to be able to connect the physics to her own experiences and she is not interested in physics 'for its own sake'. She likes interactive teaching; to discuss the content matter. She is not particularly fond of laboratory work. Her previous lack of hands-on experience with science and technology makes her feel uncomfortable in the laboratory. Therefore, when working together with a man co-student it is likely that he will be actively handling the equipment while she will passively takes on the role as secretary. She is viewed by her teachers as hardworking, diligent and neat. In the student laboratory she follows the instructions and does the prescribed assignments, instead of playing around with the equipment.

(Adapted from Hasse 2002a; Hatchell 1998; Hoffman 2002; Häussler et al. 1998; Lorenzo et al. 2006; Murphy 1991; Robertson 2006; Rosser 1995; Seymour and Hewitt 1997; Staberg 1994; Williams et al. 2003)

Also within mathematics a similar story is told; even when girls do perform well their high performance is interpreted by teachers (and sometimes also educational researchers) as a result of hard-work or rule-following rather than 'rationality' (Walkerdine 1989, 1990). Girls' good performances in mathematics are thus, according to Walkerdine, not taken as an 'evidence' for 'mathematical talent', instead they are interpreted as showing that working hard is a way of compensating for their lack of this essential 'something'.

This story about the woman students as rule-following, 'good students' and being less keen than their man counterparts to play around with the equipment is also recognisable in my empirical material, in particular among the interviews where man students' characterise their woman peers. Paul is quite typical in that he talks about the woman physics students as more disciplined and well-prepared, 'duktiga'[24] in other words:

Paul: Yeah, at least in the beginning I think it seems like the guys come there more unprepared, the girls are usually more disciplined, are more 'duktiga' kind of, they are often like that, the girls that study here.

[24] 'Duktiga' literally means capable, but it carries gendered connotations that are difficult to give justice to in translation. 'Duktig flicka' would rather translate into 'good girl'.

The latter part of Paul's utterance further reinforces the view of woman physics students as 'the exception that proves the rule'; it is possible for a woman to succeed in physics, but she needs to be outstanding. Like Henwood (1998) pointed out, such a construction of the few women in a discipline as exceptional can reinforce the notion of women in general as being unsuited for the discipline.

Lars and Bo also both talk about the woman students as being more focused on the assignment at hand, whereas the man students play around more:

Lars: Guys tinker more with stuff and girls do it properly and do what they're supposed to.

I: But, do you experience that girls and boys in the lab for example tackle it differently?

Bo: But very generally then, that... girls they are very keen to follow instructions, whereas boys dare to try things more.

Paul also talks about how the woman students are not satisfied with just getting an answer from the teacher but want to reason and discuss:

Paul: Girls like to discuss stuff more, they like to reason back and forth more than guys.

Some of the students interviewed also bring up how laboratory work may be easier for the man students as they are presumed to have more experience of tinkering, for example, with cars, Lisa when mentioning this, also talks about the woman students as more careful:

I: But if you think specifically of the lab, do you think there's any difference in how guys and girls work?

Lisa: Well, my own experience is that, not all, but many guys have an easier time with all the connections of wires and such, they have more experience with such, tinkering with cars and...

I: So, because they have the concrete experiences?

Lisa: Yes, exactly, but otherwise I think the girls are more thorough it feels like, that every step is done right.

In summary, there is an ongoing constitution of what could be characterised as a 'traditional science student femininity', characterised as being rule-following, diligent and thorough. However, these stories are not the only ones being told.

Among the often relatively careful claims about gender made by the students interviewed, Mia certainly stands out; she makes strong claims about men and women's participation in physics. For example, drawing on a discourse of gender as biological, she claimed that women are not compatible with the Discourse model of the practical physics student. But not only that, Mia also constitutes a somewhat unexpected 'science student femininity', by her explicit claim that women have more aptitude than men for logical thinking and mathematics:

> Mia: Women are more logical, women have an easier time doing logic, but that I've seen evidence of too, that girls I've had in my class or so, have had an easier time doing maths than physics.

This, I would argue, shows that there is space for negotiation of femininities and masculinities in relation to the practice of physics.

The undergraduate women I interviewed did not explicitly challenge the 'traditional science student femininity' (in contrast to the graduate student Cecilia), rather they made use of the overlap between this femininity and the Discourse model of the analytical physics student in their constitutions of physicist identities. Whereas the core aspect of the Discourse model of the analytical physics student, the physics reasoning when associated with theory, logic and mathematics, is related to the analytical technological masculinity described by Wajcman (1991), other aspects such as the structured approach, the reading of instructions, and being thorough are commonly associated with the woman students. The same goes for the physics reasoning when associated with discussions, rather than mathematical reasoning. This overlap makes it possible for the woman physics students to negotiate physics student identities that draw on aspects of the 'traditional science student femininity', but without being limited to it, also incorporating aspects of the Discourse model of the analytical physics student. I will elaborate further on how students can be interpreted as constituting science student femininities as I discuss Ann and Mia in section 9.2.3.

8.6.2 Gendering the Physicist Community and its Boundaries

In the following I focus on the gendering of two important, and interrelated, dimensions of the physicist community of practice; community and boundary. Together these define what is seen as belonging to the community of practice. As discussed in sections 8.4 and 8.5, it can, for example, have to do with the shared repertoire of physics; what is seen as important, what matters within physics. However, when defining a community it is equally important to consider what is defined as 'not belonging' in contrast to what is defined as 'belonging'; how the boundaries are defined.

I first discuss how gender can be understood as simultaneously very visible and very invisible within the physicist community of practice. Next I move on to exploring the boundary between (traditional) femininity and physics.

When considering gender in the physicist community of practice the first thing to notice is how gender is both highly invisible and highly visible within this community. The invisibility of gender in physics is reflected in, for example, an editorial by Heron and Meltzer (2005), discussed in section 2.3, where they were invited to discuss the future of physics education research:

> We highlight those directions that address intellectual issues that are specific, but not necessarily unique, to the subject matter and reasoning of physics. Therefore we omit important work on investigating gender-equity issues, for example.
>
> (Heron and Meltzer 2005, p. 390)

In their editorial they thus give gender issues no consideration at all, as they see those issues as not tied to physics as a discipline; gender is here made invisible.

The invisibility of gender in the physicist community of practice is also reflected in the Master's student Hanna's talk about physics as unusually gender neutral:

I:	Do you experience that the view on male and female is somehow different here in the physics environment, or whatever one should call it, than in other environments you spend time in or have experience of?
Hanna:	Can almost feel that here, that I, that's how I feel, that it doesn't matter so much here after all.
I:	Ok
Hanna:	That's all I've felt.
I:	What is it that matters here then?
Hanna:	It's what you know! Or, experience. It feels like it's the way you get, that you are something, it's not so much regarding gender.
…	
I:	How do you experience being a woman in the physics environment? Is it something you've reflected upon?
Hanna:	... Not so much really, now that I think about it, 'cause it's more been that I've thought that I'm young and inexperienced, it hasn't been about that I'm ... a woman. And I also think that when I look at people I have divided them like that too, not after gender but more after experience. So that actually feels good, now that I think about it. And, no, I haven't felt that anyone else has reflected over it, really.

In part I would argue that this discourse of 'gender as neutral' (Magnusson, 1998) is tied to a general academic ideal of individuality and independence. In this sense Hanna's utterance above, about how what matters is experience and knowledge, that within the physicist community of practice people are judged by what they know, not who they are, is very much in line with what Eduards (2007) argues to be the self-image of the entire academia, which, according to her, is an extremely individually oriented system:

> Because the self-image of academia is founded in a belief that here justice prevails – meritocratic, gender neutral justice – demands for equality, equal representation and democracy are seen as extraneous. The one who has the proper talent succeeds and the one that fails has primarily themselves to blame.
>
> (Eduards 2007, my translation from the original Swedish)

This individualistic focus was also expressed by the physics students in Johannsen's (2007) study of attrition in university physics, where the students who left their physics education explained their leaving in terms of an 'introspective Discourse model' (see section 5.2). Johannsen (2007) writes:

> …if students perceive that they have problems in relation to [their physics education] they interpret those problems in terms of their own perceived abilities and social identities.
>
> (Johannsen 2007, p. 145)

Thus, the students did not interpret their 'problems' in terms of, for example, reasons concerning the educational setting; what matters is you as an individual. The same goes for the 'gender as neutral' discourse; what matters is your ability and your interests, the discourse is both individualistic and meritocratic, as illustrated by the following excerpts from the interview with Tor:

I:	How come you started to study physics in the first place?
Tor:	The million dollar question.
I:	Yeah
Tor:	Have you gotten any good answers for that question? [Tor's laughing]
I:	I want your answer.
Tor:	I mean… Of course it's a difficult question, I mean it has to begin in a young age, you have to know a little bit, you have to understand mathematics in high school and… I'm not sure what the school stage is called here, but I mean 16, 17, 18, if you don't know mathematics then, you can't go into physics.
…	
I:	And you did the mathematics?
Tor:	Yes. I mean in my case, the first year of the school was open, so everybody that came into the school had to take the same classes and I guess, then it was quite obvious that I was good in mathematics and chemistry, was also a

> chemistry course, was a geology course, which I guess then shoved me into the natural sciences. And then just from that focusing the interest onto... onto what worked and physics...

This kind of indivitualistic and mericratic discourse has also been described by Erwin and Maurotto (1998) who interviewed undergraduate woman science students and found that these students generally attributed academic success purely to individual motivation, skill and luck. Further, they write:

> One of the most arresting themes that emerged from the interviews was the degree to which discourses of individualism and meritocracy shaped the world views, subjective interpretations and social practices of these young women. In these discourses the self is valorised; it is personal drive, ability, will-power and resourcefulness that shape the experiences and determine outcomes. Even when structural forces are acknowledged, they are constructed as obstacles that – with sufficient motivation and effort – can be overcome.
>
> (Erwin and Maurotto 1998, p. 63)

However, in regard to Hanna's earlier description of how physics is 'unusually gender neutral', I would like to argue that this not only reflects a general academic ideal of individuality and independence but, in important ways reflects how physics is constructed as a science that produces universal, value-neutral and objective knowledge, independent of societal factors (Schiebinger 1991). In particular, this kind of positivistic view of science has been shown to be the dominating one among undergraduate students (Ryder et al. 1999). Further, Linder and Marshall (1998) also discuss how scientific discovery most often is portrayed as a discovery of truth in physics textbooks. Thus, gender is highly *invisible* for the undergraduate physics students in terms of its intertwinement with the science as such. It should therefore come as no surprise that most of the students interviewed when talking about gender and physics do not connect gender to physics *per se.* Rather, physics was pointed out as unusually gender neutral, most likely because of its apparent detachment from societal factors. Gender is in this sense made invisible.

But gender is also invisible in a somewhat different sense. For most of the woman graduated students I interviewed, being a woman in physics is something they say they have reflected upon. Moreover, none of them seemed to find my question about 'whether they had reflected upon being a woman in physics' unexpected or out of place. The man students on the other hand met the corresponding question with surprise, laughter, or gave the impression of being threatened by it. Their positions as men in physics are something they say they have never have had to reflect upon; the physicist as a man is simply taken for granted. This, I would argue, shows the strength of the man character as the norm in physics, a norm so dominant that being a man in physics is not even something that can be talked about, a norm that pervades the community to the extent that it has become invisible.

However, gender in the physicist community of practice is also very *visible*, primarily through the high proportion of men in the discipline and how the shared repertoire of anecdotes from the history of science and pictures of scientists found in students' textbooks and on the walls of many physics departments are all largely dominated by men (see, for example, Traweek 1988).

In her interview the undergraduate student Lina talks about how this can make it difficult for woman students to constitute identities as scientists:

I: I find it interesting with gender and physics and thought we could end the interview by talking about that.

Lina: Yeah, gladly…

I: Physics is considered a traditionally male subject, is that something you've reflected upon?

Lina: Yes, very much… We talk so much about equality and that we have to get more women to apply, but I think there is very little done to really try to understand why this is the case and… 'Cause you really have to dig deep and not just use fancy language it you want to reach the problem. You can think about, how many role models do you have as a woman? You sort of have to find an identity as a scientist if you are to stay here, and then you walk around and it's portraits of men on the walls and it's male professors and female administrative staff and then it's not so easy to find an identity as a woman.

The dominance of men is something most of the students interviewed touched upon in one way or another. In fact, among the woman students interviewed, as mentioned before, Hanna was unusual in the sense that she claims to not have reflected on how she experiences being a woman in physics.

Returning to how physics is constructed as independent of societal factors, it has been claimed by Harding (1998) that the abstractness and the formality of physics need to be understood as 'distinctive cultural features, not the absence of all culture' (p. 61). Traweek (1988) has similarly characterised physics as 'a culture of no culture'. Further, when science in general and physics in particular is associated with qualities such as objectivity, reason and rationality it is simultaneously associated with masculinity as these qualities in our society are commonly denoted masculine (Benckert 1997; Brickhouse 2001). In other words, the scientific mind is at the same time viewed as masculine and disembodied, something Fox Keller (1992) has argued often makes gender-analyses of physics controversial. Often dichotomies such as active/passive or objective/subjective are used to describe and organize the world around us and in the list of words below it can be noticed how the words in the left-hand column are typically associated with physics as well as with masculinity, whereas the words in the right-hand column are associated with femininity (here adapted from Benckert 1997):

objective	subjective
active	passive
rational	irrational
mind	feeling
hard	soft
strong	weak

Benckert (1997) further discusses how both columns are not seen to carry equal value; those things that are associated with femininity are commonly seen as having less value than those things that are associated with masculinity. She also writes:

> Science and in particular physics is connected to the masculine. Words that can be used to describe physics are rational and objective and perhaps also hard. We talk about hard and soft sciences. Science is hard while the humanities are soft and within science physics is harder and biology softer and it is therefore seen as more suitable for women.
>
> (Benckert 1997, p. 59, my translation from the original Swedish)

Further, the words in the right-hand column above, the ones associated with femininity, are also associated with the 'non-scientific', thus a discontinuity is created between femininity and science. This discontinuity between femininity and science is one that Schiebinger (1991) argues can be traced historically; as masculine/feminine and scientific are not static conceptions, but living ones that have been defined and redefined throughout history. As was introduced in section 4.5, science has, according to Schiebinger, been defined and made academic through both the exclusion of individual women, and the exclusion of areas that have been traditionally dominated by women (e.g. health-food) and of ways of doing science that have been accessible to women (e.g. family-based laboratories). By these processes, what is seen as feminine, and what is seen as scientific have been defined in opposition to one another; what is scientifically correct is automatically unfeminine and characteristics seen as feminine are by the same token undesirable in science. Berner (2004) discusses how science's ideal of rationality (an abstract, disembodied and distanced thinking) historically has been contrasted to women's embodiment and intuition and their presumed lack of rationality.

This discontinuity between femininity and physics, this boundary between the physicist community of practice and the doing of 'traditional femininity', does in very real ways affect the identity constitutions of, in particular, some of the woman students I interviewed. These women can be interpreted as making use of this perceived boundary in two different ways. Karin and Marina both, in different ways, acknowledge that they, as women doing physics, break the norms of physics. Ann, Lisa and Cecilia, on the other hand, all position themselves as being different from other women: as non-participants in a 'traditional femininity', a non-participation that, then, in turn, makes a participation in physics possible. I am here not making the argument that there is such a thing as a 'traditional femininity', in the sense of a community of practice where the participants are taking part in well-defined activities. However, for the students who define themselves in opposition to such a 'traditional femininity', the idea of such a community nonetheless functions as contrast to both the physicist community of practice and their own doing of gender. By a rejection of the feminine these women are therefore able to claim membership in the physicist community of practice. This is well in line with how Paechter (2007a) in her research on tomboys has found that they, in their identity constitution, 'reject more or less the same things as boys reject when they construct femininity as Other to the man subject' (p. 19). Paechter concludes 'there seems to be no way of claiming a powerful position without embracing a masculinity which also appears to involve a repudiation of the feminine' (p. 20). This denial of the feminine is often understood as something 'negative'; as women having to give up something of themselves to fit into a masculine environment (see, for example, Kvande 1999). While this can be true for some women, I would argue that it is an oversimplification to always make this interpretation. In the case of the PhD student Cecilia, for example, an alternative interpretation could be that by choosing physics she is able to perform a masculinity – not all women desire to be seen as feminine. The choice of physics, then, also in itself, becomes a part of a doing of masculinity for Cecilia.

8.7 Concluding remarks

In this chapter I have approached the interviewees as a collective group, with the aim of exploring how a physicist community of practice is constituted, with a particular focus on the gendering of this constitution. This collective analysis is built on the previous chapter's narratives of the individual students, but in this chapter the level of abstraction has been raised. This was accomplished through the incorporation of the analytical tool of Discourse models and, in the latter part of the chapter, by a utilization of the theoretical framework. As an empirical complement and comparison, some of the interviewees not presented in Chapter 7 have also been introduced.

In the next chapter the analytical level of abstraction is raised further, as the full diversity of the conceptual framework is used to deepen the exploration of how students constitute the practice of physics and their own identities as physicists, with a particular focus on their simultaneous doing of physics and doing of gender.

CHAPTER 9
Identity in Practice: Becoming a Physicist?

9.1 Introduction

In the previous chapter I approached the interviewees as a collective group; using the students' collected accounts to portray a physicist community of practice. Now I turn back to the individual interviews. In the following chapter I will theoretically deepen the exploration of how the students are constituting identities by simultaneously applying the analytical tools of positioning and Discourse models with my theoretical framework, to the student narratives. In doing so, the full complexity of the conceptual framework is utilized in order to explore how physicist identities are constituted. Some of the students are paired to make it possible to contrast and compare how they are constituting their identities. Thus, I have chosen to structure the chapter in a way that is closely tied to the individual students. The analysis in this chapter will include me drawing on the analytical content and its transcript exemplars that I have already given in previous chapters but without explicitly cross-referencing them. At the same time, it is important to point out that this chapter is more about understanding possible ways of participating in physics than understanding the individuals. For example, which positionings are possible and which are seen as desirable?

The students in this chapter are chosen because they represent a wide array of possible ways of engaging in the practice of physics. A more thematic structure could also have been possible, but while the complexity of the practice could have been also captured in this way, the complexity of how the individual students are negotiating their identity constitutions would most likely have been lost.

The starting point of this chapter is an understanding of identity as a negotiated and negotiating experience, a work that is always going on, something that we do ourselves, as well as something that is done to us. A key concept in this understanding of identity is 'participation', how we, through participation (and non-participation for that matter) in various communities of practice, constitute our identities. In relation to the students' doing of gender it can be seen how gender is more explicitly in focus among the woman students, even though the man students' identity constitution can be interpreted as participations in different (partly class-based) masculinities.

The chapter is divided into two parts, where the undergraduate students are treated in the first part and the graduate students in the latter.

9.2 The Undergraduate Students

In the following section I return to the undergraduate students that were introduced in Chapter 7. Here the level of abstraction is once again raised; the analytical tools of positionings and Discourse models as well as the theoretical framework are employed to explore and problematise the undergraduate students' constitutions of identities and of the practice of physics. In doing so I also draw on the analytical outcomes from Chapter 8.

9.2.1 Kalle and Paul – the Laboratory as an Arena for Tinkering

The tension between the Discourse models of the analytical and the practical physics students are prominent in how both Kalle and Paul constitute their identities. Taking pride in their practical skills and having somewhat similar experiences from working-class jobs they both associate themselves with the Discourse model of the practical physics student. But, while for Kalle the identification with this Discourse model is relatively unproblematic, even though there is some doubt about what 'really' counts as physics, Paul's constitution of a physicist identity is more of a struggle. In her classic study 'Gender and subject in higher education' Thomas (1990) writes:

> For these men, the physicists and the physical scientists, self-image was concerned with the particular nature of one's abilities and the sort of job one could do with them; it was rarely concerned with one's identity as a *man*, as opposed to scientist: the two were one and the same thing. There was, as we might expect, no conflict between the two.
>
> (Thomas 1990, p. 116)

The relationship between men/masculinity and physics is often depicted like this; as unproblematic and straightforward. However, in the case of Paul, I would argue that the tension he experienced between the Discourse models can be interpreted in terms of a conflict between different, partly class-based, masculinities. Paul comes from a background where practical skills are highly valued; Paul sees himself as someone who learns by doing and is proud of his ability to quickly master a new job. His practical skills and hands-on approach to learning is also something he brings into the student laboratory. This is well in line with the practical masculinity described by Wajcman (1991), and Paul would likely recognise himself in how the Swedish engineers studied by Mellström (1995) describe technical skills as an important part of what it means to be a competent man. I would therefore argue that the doing of physics that Paul was engaged in in the beginning of his physics education – the association with the Discourse model of the practical physics student – can be interpreted as a doing of a particular masculinity. Despite Paul's personal valuing of practical skills, he does, however, recognise that these skills may not be so highly valued in the physicist community of practice and as a consequence Paul is trying to align himself more and more with the Discourse model of the analytical physics student. By doing so Paul is aspiring to be on an inbound trajectory in the physicist community of practice. All in all, Paul is unusually able to articulate how he is continuously working with his identity; his struggle to combine different (gendered) ways of being a physicist into the nexus of multimembership that is his identity. As in the earlier cited description from Thomas (1990), there is for Paul perhaps no conflict between being a man and being a scientist, but rather between *how* to be a man and *how* to be a scientist.

In the case of Kalle there are also certain tensions between his preferred way of doing physics and what he experiences to be the norms of the physicist community. Kalle does experience a continuity between the physicist community and the workshop community he has previously been a member of; he values skills developed in the workshop community and uses the workshop as a metaphor for working in a physics laboratory. For Kalle, the workshop could be seen to function as a boundary object that creates a continuity between his previous experiences and his experiences of the student laboratory. Further, the strength of the metaphor lies in its capacity to allow one to understand and experience one thing in terms of another (Lakoff and Johnson 1980), thus, by thinking about the physics laboratory in terms of the, for him, well-known workshop, he is able to interpret the laboratory practices in terms of the practices of a workshop.

However, Kalle does experience that the physicist community of practice does not fully appreciate his highly valued practical skills; as with Paul, there is a conflict between the masculinity valued in his background of industrial work and the physicist community of practice. What attracted Kalle to physics in the first place were the experimental, practical aspects of the discipline. This is a fascination that still influences his engagement in the practice of laboratory work, and that, according to him, sets him aside from the 'ordinary scientist'. Further, Kalle's continued identification with what can be interpreted as a marginalised masculinity in physics keeps him on a peripheral trajectory in the university-based physicist community of practice and makes him unwilling to stay in that community. Thus, unlike Paul who is explicitly working on aligning his engagement in the practice of physics with what he experiences to be a higher valued form of engagement, Kalle is on an outbound trajectory towards work in industry, a community of practice that does share Kalle's valuing of practical skills.

In Paul's narrative there is a very explicit sense of how the work of identity is ongoing, how he is renegotiating both his way of practicing physics (as discussed above) and what it means to be a physicist. Thus, Paul's identity constitution is much more complex than just the passive transformation into a fixed physicist identity, what Hildebrand (2001) criticised situated learning theory for. For example, in the interview excerpt below Paul is renegotiating what it means to be a physicist:

> Paul: No, I… it would be fun, but that I should become a researcher, that's not something I thought. I've grown up with pals who are everything from construction workers to… yeah, everything, but there are very few people who've continued to higher education among my friends. And, researcher, yeah, it would be fun, but I've never thought of myself as smart enough for that, but since I started here I've realised that maybe it's not about being so very smart, but it's about thinking that it's fun, to be interested, and have discipline.

This could certainly be read as Paul coming closer to realising what physicists are 'actually' like: that they are not necessarily extremely smart, but that important characteristics for a researcher include interest and discipline. Nonetheless, in making this explicit, Paul is for himself negotiating what it means to be a physicist. In doing so, he, as a 'legitimate peripheral participant' in the university-based physicist community, is also shaping the meanings that matter within that community.

In summary, an important part of the learning to become physicists for both Kalle and Paul are their struggles to define the enterprise and to reconcile conflicting interpretations both of what is means to be a man and what it means to be a physicist.

9.2.2 David and Susan – Analysis is Everything

In contrast to Kalle and Paul, David's and Susan's constitutions of physics student identities are characterised by an association with the Discourse model of the analytical physics student. But, not only do Susan and David see the practices associated with this Discourse model as most suitable for them, they also see them as the most appropriate ones in a physics student laboratory. Both of them thus constitute the boundaries of the practice of physics as enacted in the student laboratory very much in terms of a simultaneous inclusion of the Discourse model of the analytical physics and exclusion of the Discourse model of the practical physics student. Thus, for them, the shared repertoire of the physics student laboratory includes aspects such as a focus on the analysis, reading of instructions and being well-structured. In particular David is very adamant that practical skills are not important in the student laboratory and that the tinkering approach to laboratory work associated with the practical technological masculinity is not an appropriate one. His constitution of an identity as a physics student is therefore characterised by a distancing from what he experiences in this context as an unproductive masculinity. Further, he also portrays the woman physics students as being closer aligned with the more successful Discourse model of the analytical physics student.

In school, David found it difficult to identify with the practically oriented physics, but at 'basåret'[25] he met a different, more mathematically oriented physics more in line with his own interests. Susan also had bad experiences of laboratory work in school and she too came to university with an interest triggered by the theoretical side of physics. In a sense David's and Susan's somewhat narrow trajectories, confined within the Discourse model of the analytical physics student, are not completely dissimilar from Kalle's strict association with the Discourse model of the practical physics student. However, whereas Kalle (and even more so, Paul) struggle to reconcile 'their' way of doing physics (and doing masculinity) with the dominant discourse of physics, no such reconciliation seems to be necessary for David and Susan – at least not in relation to the practice of the student laboratory. David does not even recognize the existence of competing Discourse models, but claims that his way of doing physics is 'the' way of doing physics, shared in general by his course-mates. Thus, David is constituting himself and the physics norm in close interaction.

However, despite David's explicit distancing from the Discourse model of the practical physics student in the practice of the student laboratory, he does recognize the importance of practical skills in the research laboratory. Thus, for David there exists a discontinuity between the practice of the student laboratory and the practice of the research laboratory. This perceived discontinuity could in part be due to the fact that David's engagement in the student laboratory has made it possible for him to negotiate the meaning of that practice for himself, while the practice of the research laboratory is something, in a sense, beyond his reach in terms of engagement. He can of course imagine it, though, but this imagination is possibly based on a somewhat stereotypical assumption about the practice of a physics research laboratory as being associated with the practical technological masculinity that David so strongly distances himself from.

[25] A one year compensatory education for those who want to study science, technology or medicine at university but did not do the science stream in secondary school.

9.2.3 Mia and Ann – 'the Anomaly of a Woman in Physics'[26]

In both Mia's and Ann's[27] identity constitutions the relation between femininities and the participation in the physicist community of practice is central. Further, they both constitute their physics student identities largely by an association with the Discourse model of the analytical physics student. Unlike David and Susan they do not, however, experience the boundaries of the physicist community of practice as coinciding with this Discourse model. On the contrary, both Mia and Ann do value practical skills in the student laboratory, even though they do not see themselves as possessing such skills in the context of doing laboratory work.

When Ann talked about how she worked in the student laboratory she said that she was rather passive and let her man laboratory partner handle the laboratory equipment. This passivity is expressed by her as 'taking the female role' (see section 7.2.6). Ann says that the reason she took on this 'female role', is that she 'felt a bit slow'. Thus Ann associates 'the female role' with 'being slow' and passive: hardly sought after characteristics in science, thereby constituting a boundary between science and femininity. As discussed in sections 4.3.2 and 8.6.2 this is a discontinuity that can be traced historically, as what is seen as feminine and what is seen as proper science ever since the Enlightenment have been defined in opposition (see, for example, Brickhouse 2001). For example, feelings have been defined as simultaneously feminine and unscientific.

Further, Ann's use of the expression 'taking the female role' shows how femininity for her is something she does, rather than something she is. Mia, on the other hand, draws on a discourse of gender as something biologically determined in her talk about gender and physics and is, from that standpoint, very explicitly expressing a discontinuity between femininity and the participation in the physicist community of practice:

I: It's mainly men who study physics – is that something you reflected upon? Why it's like that..?

Mia: Yeah, that I also believe is, depends on the brain, on the woman and the man brain. That it's easier for men, that's the way it is... I did the science programme in high school too, and the women were a minority and that's the way it is in most classes, or all I know about, so that's probably something that has to do with guys having an easier time with physics.

The discontinuity Mia experiences between femininity and physics is possibly the most obvious in the laboratory setting, where it, according to Mia,

[26] Title of an essay by Fox Keller (1977).

[27] The interview with Ann was used in Chapter 4 to illuminate the theoretical framework, and certain excerpts and analytical outcomes from that chapter are repeated here. This is because I want Part III of the thesis to be able to be read relatively independently from the rest of the thesis.

gave rise to a particular division of labour between her and her man laboratory partner; he was handling the equipment whereas she was handling the computer. This discussion about their division of labour took place prior to my explicit introduction of issues of gender and physics into the interview. When asked why they had this division of labour Mia says that it is 'in the genes', thereby referring to a biological discourse about gender.

Further, Mia is in her constitution of her identity simultaneously reinforcing a boundary between physics and mathematics, a boundary that according to her is gendered. Women have, according to her, more aptitude for logics and mathematics whereas men have more aptitude for what she sees as 'physics'. She also frames her own participation in science in terms of a participation in such a localized femininity, characterised by an aptitude for logical thinking and mathematics:

In summary, Mia, thus, simultaneously constitutes the practice of physics as gendered and her participation in it, her identity as a physics student, as gendered. And, her participation in a certain localised femininity does in very real way shape her participation in the physicist community of practice, in terms of, for example, how she engages with other participants.

Even though Ann at times has taken up the 'female role' in the student laboratory, it has been primarily from a different perspective; that the perceived discontinuity between femininity and physics shapes her identity constitution. In the interview Ann is positioning herself as different from 'normal' women and repeatedly comes back to how she is more 'comfortable among guys' than in woman-dominated environments. Thus, Ann can be understood as positioning herself as a non-participant in 'traditional femininity'; a positioning that makes a simultaneous positioning as a participant in physics possible. As discussed earlier in sections 8.6.1 and 8.6.2, such a positioning as a non-participant in traditional femininity is well-documented in earlier research, which has shown that women within science and technology often construct them themselves as 'one of the boys' in order to be able to position themselves as scientists/engineers.

In their interviews both Ann and Mia are positioning themselves as associated with the Discourse model of the analytical physics student, or more precisely by negotiating a participation in physics that draws on characteristics that are simultaneously associated with femininity and with this Discourse model. Mia describes a localised femininity that she associates with an aptitude for logic and mathematics. Ann refers to characteristics that are commonly associated with woman science students, such as being well-prepared and wanting to reason and discuss. For example, throughout her interview Ann returns to how it is the discussions about the laboratory exercise, the prolonged engagement with the content, that she finds useful, not the laboratory work as such. Thus, both Ann and Mia are negotiating a simultaneous participation in a femininity and in science, which I, in section 8.6.1.2, discussed in terms of 'science student femininities'.

In the interview with Ann the complexity of 'doing gender' in the participation in physics is very explicit; she describes herself as taking on 'the female role' in the laboratory and describes her practice in the laboratory in terms of a rather traditional science student femininity. Ann does this while also positioning herself as being different from 'normal women'. This complexity of Ann's identity constitution is further reinforced as she, in the context of doing maintenance work on her car, is positioning herself as a practical person, despite her claims of her being unskilled in practical work in the student laboratory. This shows how subject positions are most certainly highly context-dependent and how they vary between situations. Further, Ann is also distancing herself from a particular kind of technological masculinity, represented both by the men at her earlier workplace and by her course-mate Mats. By distancing herself and physics from the kind of masculinity that characterised her previous workplace, the electrical workshop Ann is constituting a discontinuity between communities that are very similar to those that Kalle constitutes a continuity between (see section 9.2.1). Ann further describes her previous workplace as very 'laddish' and is letting this workplace represent 'masculinity'. Thus, for Ann, the somewhat less man-dominated physics does not come across as particularly masculine, thereby demonstrating the relational nature of gender. However, Ann stresses how physics for her is a 'very open' community, wherein a multitude of different people and different ways of participation can be included. For example, as discussed in section 4.4, she is positioning her course-mate Mats (who represents the Discourse model of the practical physics student) as 'fitting' in the student laboratory:

> Ann: Mats, he's an experimentalist! He's so much fun to do labwork with, 'cause he really gets, he might not understand the theory at all and hasn't done anything and is tired and hasn't slept and he sure starts to tinker kind of! He's so very different [from me], he really fits in a lab!

This shows how multifaceted and complex the practice of the student laboratory is. But, this interview excerpt is also interesting because of the different possible interpretations that can be given to it. How Ann turns Mats' lack of preparations and poor theoretical background, which could easily have been viewed as inappropriate in a student laboratory, into something positive, can be understood in at least two different and possibly complementary ways. Firstly, Ann's positioning of Mats as someone who fits in a laboratory is most certainly made possible by Mats being a man, and as such, fulfilling the gender-norms of physics. This makes it possible for Ann to interpret Mats' possible 'shortcomings' as desirable, or at least suitable, qualities in physics. But, secondly, Mats' doing of a practical, technological masculinity is also a 'doing of class', one that Ann not only recognises from her previous experience of working in a workshop, but also values, and to a certain extent identifies with. Thus, in order to understand Ann's positioning of Mats it is important to take both gender and class into account.

9.2.4 Lisa – the 'Otherness' of Age

Lisa associates herself with the Discourse model of the analytical physics student; the primary goal of laboratory work for her is learning of theory. According to her, a good laboratory practical is one that is able to yield results as close as possible to the ones expected from theory, and she is thus affirming a dominant discourse of scientific certainty. She can also be understood as constituting her identity as a physics student very much in terms of a participation in a 'traditional science student femininity'. Overall, issues of gender are important in her identity constitution (though perhaps not to the extent as for Mia and Ann), but here I am going to focus on one particular aspect, the interplay between gender and age. Lisa started studying relatively late, after having spent a few years working and travelling. This 'break', she says, gave her the opportunity to really think about what she wanted to study, and she chose astronomy out of interest instead of 'just picking some education that seemed good'. This implies that astronomy might not have been an option for her had she started studying earlier.

When I ask her about why she thinks that more men than women study physics, she says:

Lisa:	I think it's something ... that they've gotten that interest more stimulated as kids than the girls have.
I:	Ok.
Lisa:	Technology and such.
I:	What guys are supposed to do and so on. So why are you here?
Lisa:	I don't know [laughter] I'm not like everyone else, I walk my own road I think. I did start studying really late.

Here Lisa is firstly constructing men and women as different, due to a difference in socialisation. This construction is clearly a co-construction with me, the interviewer; I asked my question about men and women and an answer reinforcing this difference is what is to be expected. But, then something interesting happens; when Lisa is asked to explain her own choice of physics she does this by constituting herself as different from other women. In doing so she is utilizing the same discontinuity between traditional femininity and physics as Ann is, but from a somewhat different perspective. By explicitly referring to a time aspect, that she started studying late, Lisa is drawing on an 'otherness' of age. For Lisa, then, her age adds the same kind of complexity to her identity constitution that Hughes (2001) described in the case of ethnicity; one of the woman science students in Hughes' study was able to constitute a comfortable identity as a scientist by drawing on a discourse of

Asian minority students as highly motivated. By drawing on an otherness (in terms of age or ethnicity) both Lisa and the student in Hughes' study are able to constitute comfortable identities as woman scientists, without challenging the masculine norms of science. To summarize, in order to understand Lisa's constituting of her identity it is important to take both gender and age into account.

9.3 The Graduate Students

So far in the chapter the practice of the student laboratory and the undergraduate students' associated identity constitutions have been addressed. Next I turn to the graduate students, for a similar exploration of the research laboratory. The difference is that here I also focus on their constitutions of differences and similarities between the student and research laboratory, as well as on their constitutions of boundaries of the physicist community of practice.

9.3.1 Cecilia and Tor: The Experimentalists

Cecilia and Tor both strongly identify, not only as physicists, but more exactly as experimental physicists. Central to both of their participations in the practice of the student laboratory is an unproblematic association with the Discourse model of the practical physics student. But whereas Tor describes his practice more towards the professional end of the spectrum, Cecilia's approach was more of the 'playful' kind. He talks about learning to handle equipment, in a progression from simple instruments like voltmeters to more advanced ones; she talks about playing around with the equipment until it works. Both thus focus extensively on the handling of equipment in their description of their practice in the student laboratory, but from different perspectives. Despite Tor's and Cecilia's strong association with practical Discourse models it can also be noticed how aspects of analytical Discourse models are present in their talk. This is, however, expressed without the kind of tension that is apparent in Paul's and Kalle's identity constitutions (see section 9.2.1). Tor talks about how it was an aptitude for mathematics that attracted him to the natural sciences in the first place, and, more noticeably, when Cecilia is asked to characterise the physicist this is done very much in terms of the Discourse model of the analytical physicist. Furthermore, Cecilia confidently answers 'yes' to my question about whether she recognizes herself in the attributes she associates with 'the good physicist'; being analytical, logical and creative. She does not however seem to notice that her description of 'the physicist' differs markedly from her description of her own practice as a physicist. Both Cecilia and Tor are thus very explicitly bringing out aspects of both physics student Discourse models, thereby constituting their physicist practice as a highly versatile one – an association with the physicist Discourse model.

Cecilia's willingness to describe the physicist can further be interpreted as her experiencing herself to be in a position where she has legitimacy to describe, and thereby also define, norms of the physicist community of practice. In other words, the identification with and the negotiation of the physicist community of practice are tightly interconnected.

While for some students it is difficult to gain access to a position as a legitimate peripheral participant in the physicist community, this never seems to have been the case for Tor. He experiences a strong continuity between the student laboratory and the research laboratory, seeing the difference between the two in quantitative rather than qualitative terms. Tor, at least in retrospect, perceives the practice of the student laboratory as authentic. In a sense, the continuity between the student laboratory and the research laboratory can be understood as working both ways for Tor; not only does he see the practices in the student laboratory as authentic, but he also talks about an ongoing learning process, how doing research is about learning:

> I: When do you ask for help in your studies? In your research?
>
> Tor: Very often.
>
> I: For what reasons?
>
> Tor: There is a lot I don't know. There is a lot I can't know. There is a lot I can't possibly know.

In their interviews Tor and Cecilia are both positioning themselves as (experimental) physicists; they both value practical skills and Cecilia talks about how what attracted her to experimental physics was its connection to 'reality'. Furthermore, Tor expresses great pride in being able to handle an advanced experimental set-up without much knowledge of the theoretical background. In doing so he signals a view of the execution of the experiment as more fundamental than the associated theoretical considerations. Here, it can be noticed that it is not experimental physics in general that Tor so highly values, it is experimental physics that in a sense gives direct access to the studied phenomena:

> Tor: And, I mean, a lot of experimental physics is boring, you just get some really ugly graphs, you just don't understand and stuff like that.

Thus, for Tor, the most highly valued kind of physics is a kind that makes the layer of theoretical interpretations between the experimental outcomes and the researcher redundant. For example, Tor talks with exaltation about his construction of a scanning tunnelling microscope; how he created a machine that could take pictures of atoms! About this experiment he says:

Tor: ...it can't get more fascinating than that. That is just a simple fact.

Cecilia's 'playful' approach to doing physics in the physics laboratory is also interesting from a different perspective – how it is intertwined with her doing of gender. In Hasse's (2002) anthropological study of Danish physics students she found that a playful approach to physics in general and laboratory exercises in particular was something that was limited to a group of man students, none of the woman students in her study participated in this playing activity. While Hasse's results of course cannot be claimed to be 'generalizable' to a different context, it is still noticeable how Cecilia's approach to laboratory work is so very different from all the women in Hasse's study. Further, as shown earlier (see section 8.6.1.2) woman physics students are often interpreted as being diligent, neat and closely following the prescribed assignments. This is a picture that is far from Cecilia's description of her participation in physics, in particular in terms of her practice in the student laboratory. In fact, I would argue that Cecilia's talk about her practice in the student laboratory as being characterised by inadequate preparation and unafraid tinkering with the equipment, in part needs to be understood as a reaction towards the norm for woman science students. Thus, in her identity constitution as a woman in physics Cecilia not only has to relate to the masculine norms of the physicist community of practice and the norms of 'traditional femininity', but also to the expectations there are on woman physics students.

But not only is Cecilia rejecting the science student femininity, she is also rejecting a traditional femininity in more general terms, for example in the following interview excerpt:

I: Before you talked about yourself a bit as a 'teknolog'[28], but how would you define yourself?

Cecilia: Erm ... kind of ... Partly I'm sort of a 'teknolog', but I come from [city] where the science students are like 'teknologer' so to speak, sure run around in overalls and ... drink beer, if you see that as, and sing indecent songs – so, yes, then I'm a 'teknolog' [both of us are laughing]. But ... at the same time I like to read books and am interested in politics and such things, and that might not be so 'teknologigt'.

This rejection of traditional femininity is also explicitly expressed through her description of herself as 'laddish' (see section 7.3.1). On the one hand, this non-participation in traditional femininity serves the same purpose as for Ann (see section 9.2.3); by drawing on a discontinuity between traditional femininity and physics they are able to position themselves as non-participating in one and participating in the other. But not only that; through

[28] 'Teknolog' is colloquial Swedish for 'engineering student', carrying a notion of participation in engineering student traditions.

her participation in physics Cecilia is also able to participate in a particular kind of masculinity. Thus, by her participation in physics she simultaneously constitutes her own gendered physicist identity and the community of practice as gendered. This may not challenge the masculinised gender norms of physics, but it most certainly challenges the norms for woman physics students. Finally, it should be noted that Cecilia's rejection of traditional femininity is not a rejection of gender as important in the physicist community of practice. Cecilia is reflective and well versed about gender issues. This is in clear opposition to, for example, Hanna who describes having not reflected on being a woman in physics and in a sense goes against my positioning of her as a woman in this context; claiming that what matters in the physicist community of practice is experience, not gender. Thus, Hanna can be interpreted as trying to uphold a gender-neutral position.

9.3.2 Klara and Hanna: Becoming a Physicist – or not?

Klara and Hanna are both Master's students in the same research group, working with somewhat similar projects. However, their educational backgrounds are quite different. Klara is enrolled in the Engineering Physics programme, a programme Hanna decided against because she perceived it as too difficult and was unsure if she could cope with it. Instead Hanna chose a more interdisciplinary engineering programme, also motivated by its possibilities to combine mathematics and physics with broader societal perspectives.

It can be seen from the interview that despite the many similarities of their Master's research projects, Klara and Hanna are positioning themselves very differently. Klara is currently positioning herself as a physicist, a positioning that was largely made available to her through her prolonged engagement in a community where 'everyone' is a physicist:

I:	Do you see yourself as a physicist?
Klara:	I think I've started to do it, now.
I:	When did you start doing it?
Klara:	Well, yes. After a while when I had started the Master's research project.
I:	Ok.
Klara:	You are ... We have been a group where all are physicists and that ... We are physicists. So then I started, I'm probably a physicist too. [we're both laughing]

Hanna, on the other hand, is not positioning herself as a physicist. In her interview she explains that for her to gain access to the physicist community of practice not only a PhD in physics is needed, but also some years of working experience after that. Hanna's lack of identification with the physicist community of practice is also intertwined with a perceived inability to negotiate what it means to be a physicist. She says that she finds it difficult to say much about the characteristics or experiences that are useful for a physicist (or an engineer):

> I: What is important to be skilled at then, as a physicist or engineer or whatever you're on your way towards?
>
> Hanna: ... Well, it's kind of to be able to like be independent or ... be, critically examining ... I don't know. Erm.

So, from Hanna's perspective the boundaries of the physicist community of practice are tightly set and gaining access and legitimacy is difficult. Further, from her perspective the boundaries are diffuse, and perhaps also moving, being defined by 'experience', rather than being defined by an explicit marker of membership, such as a degree.

In relation to doing laboratory work as an undergraduate student Klara talks about herself in terms of the Discourse model of the analytical physics student; she was not very interested in doing experiments but describes herself as having more aptitude for theory. During her Master's research project she describes this as having changed markedly; she says that she currently has much more self-confidence in the laboratory, as she during her Master's research project has had the opportunity to try things herself, without the pressure of time she experienced in the student laboratory. Consequently, during her Master's research project Klara could be said to have increased her repertoire of skills in the laboratory to include aspects of both the physics student Discourse models, thereby also increasing her repertoire of possible ways of participating in the laboratory practice. In other words, she has increased her versatility; she is able to associate herself more with the physicist Discourse model. Interesting to note, however, is that she makes no evaluation of the different physics student Discourse models; unlike for example David (see section 7.2.3), she makes no claims about whether she perceives one of the Discourse models as more highly valued. Her distancing herself from the Discourse model of the practical physics student is explained not by a view of 'handling of equipment' as lesser valued than 'handling of theories', but because she did not feel comfortable with the laboratory equipment. Further, Klara points to the time aspect as one of the main differences between the student laboratory and the research laboratory; in the student laboratory time was always sparse and she says that she was not able to develop the kind of independence she currently experiences in the research laboratory. Klara also talks about a certain continuity between the student laboratory and the research laboratory, in terms of the student laboratory helping her to get accustomed to certain kinds of equipment, but there are also prominent differences:

I:	Is there a difference in the purpose in the laboratory you're working in now versus in the student laboratory? Or maybe rather, what are the differences?
Klara:	Yes, there's a difference. Now I'm studying something to see what I get kind of, but then it was, then it should ... Yeah, what should I say. Then it wasn't exploring in the same way, then it was more about confirming what we already knew. And now it's more research. So to speak.

Overall, there is a clear sense of progress in Klara's narrative. She was attracted to the Engineering Physics programme because of a rather idealistic interest in space research, an interest that along the way has transformed into a more focused aspiration to work with applications of radiation. Further, while recognising a certain continuity between the student laboratory and the research laboratory, she is also reflective about the differences and about her change from student to physicist and her associated increased repertoire in the laboratory. All in all, Klara could thus be understood as being on an inbound trajectory in the physicist community of practice.

Hanna also describes her approach to laboratory work as an undergraduate in terms of the Discourse model of the analytical physics student, with a particular focus on thoroughness. Thus, Hanna's approach to laboratory work can be described as being located in the overlap between the Discourse model of the analytical physicist and the 'traditional science student femininity'. Noticeable in Hanna's talk about working in the student laboratory versus working with her Master's research project is how little distinction she makes between the practices; when asked directly she does mention that the Master's research project is less premeditated than exercises in the student laboratory, but the clear sense of progress that characterises Klara's narrative is lacking. When you are doing your Master's research project you are 'nothing' Hanna says, neither physicist nor engineer or anything else. When Hanna started her Master's research project she thought of it as 'simple', something anyone could do:

I:	Do you experience ownership over your project? Over what you do?
Hanna:	... Erm ... Maybe a bit more now towards the end.
I:	Yes.
Hanna:	When one starts to realize that it's me who has done this and ... there is no one else that knows what I've done, almost. So, then you get to do it...
I:	What has changed during the course of the Master's research project then?
Hanna:	Mmm... It might be that to begin with I thought that what I'm doing is so very simple and...
I:	What do you mean by that?
Hanna:	Well, easy. I sit here and simulate small things.
I:	Aha
Hanna:	And anyone could do that and it doesn't take much to do that. And, I also felt that I'm not doing much, I've been here for so long and I haven't done much,

some how. But then, and that might be because I came straight from school, where everything is supposed to go so quick and ... then you start to realize that maybe it's like, I've been working all the time, so it's not like I'm wasting my time...

I: No, no.

Hanna: ...maybe it doesn't go faster than this. And, ok, it might not be the most difficult thinks I've done, but ... the conclusions and so on one has made from it, it's still, are valuable after all.

Reflecting about the Master's research project in the interview situation she does, with hesitation, voice how she, after all, has come to made 'valuable' contributions. However, the contrast to Klara is striking. Whereas Klara's clear sense of progress has opened up an inbound trajectory in the physicist community of practice I would argue that Hanna, at most, can be understood as being on a peripheral trajectory.

9.3.3 Karin: On the Boundaries of Physics

Among the graduate students interviewed Karin stands out as the only one not working with a well-defined research project. Even though the freedom and interdisciplinarity of her project is something that she says attracted her to the PhD project in the first place, this also gives rise to tensions in a PhD student community where well-defined projects with clear outcomes are the norm. For example, Karin experiences not having such a well-defined project as stressful; she desires more guidance and wants to see concrete outcomes in terms of publications:

Karin: And, that, especially now in the beginning I would want so much more concrete guidance kind of, but, I don't think that I have a project yet. Well, I think about a project as something that can result in a publication or an article.

Thus, there is a tension for Karin between wanting independence and having too much freedom. The interdisciplinarity of the project also gives rise to tensions concerning how well her project fits in at the department:

Karin: So it's not as, erm, what should I say, as advances, or not as technological or as much on a detail-level perhaps as others who are doing their PhDs in solid state physics.

On the same note, Karin also talks about lacking colleagues to talk about her project with, in fact she even finds it difficult to answer my question about how many colleagues she has got, as she does not view herself as really belonging to any research group. Thus, in terms of her research area Karin experiences herself to be on the boundaries of the physicist community. For Karin the physicist is defined by having a strong and specialised interest on the 'detail-level' – like her colleague Ann who is interested in 'band gaps and electrons jumping' – not her own more interdisciplinary interests. In other words, Karin is lacking the 'nerd component' of the physicist Discourse model.

This difficulty – or perhaps reluctance, rather – to 'fit in' is something she relates as having struggled with since the beginning of her education. She had doubts about choosing an engineering education, as she perceived it to be only for 'the smart' people, whereas she saw herself as more 'ambitious' than 'smart':

> Karin: Because, in secondary school then, my thought then about engineers was kind of that it's just the smart ones that study it. And I had, I had very good grades in secondary school because I was very ambitious.

This is well in line with how many women tend to 'explain' their high grades and could be read as Karin participating in a traditional science student femininity that implies if not a non-participation in the physicist community of practice, at least a peripheral participation:

> Karin: And ... I'd like to be good at what I'm doing and then if it's not considered the most difficult part of physics, then, it's OK kind of, but I want to be good within my field and I want to, I want to be good at this.

It was also very important for Karin to point out that when she eventually chose to do an engineering education this was because this education was the one that best fitted her interests in applied science and mathematics, not because it was a prestigious education. Thus, from the start Karin has had a hard time identifying with 'the engineer' and, now later, with 'the physicist'.

Consequently, Karin talks about herself as a 'PhD student' and not as a physicist. She says that 'she is still learning to learn', something that for her is associated with being a student, not a professional, and that for her defines a discontinuity between the physics student and the physicist. This can be contrasted to how Tor perceives continuous learning as a common element in student and professional physicist communities and thus constitutes a continuity between the two.

All in all, from Karin's perspective the requirements for becoming a participant in the physicist community of practice are quite strict; being smart, having a very narrow and specialised interest, doing 'advanced, technical' research and a highly specialised education. In addition, the boundaries of the community is gendered in the sense that Karin spontaneously associates physicists with 'older men' – an association she acknowledges to be 'sad' – but still contributes to her own feeling of alienation. In summary, I would argue that Karin's perceived non-participation in the physicist community of practice must both be seen in relation to her participation in a rather traditional science student femininity and the peripheral position of her research area. However, here it also needs to be noticed that the physicist community of practice not is not necessarily the most desirable for Karin. In contrast to Hanna, who finds it difficult to present an alternative to the physicist identity she rejects, Karin does this, at least to some extent, by drawing on the interdisciplinarity of her research area and of her education:

> Karin: Erm, 'cause I don't think that I've studied, well I've studied physics, but not as a physicist, feels narrower. I've studied so many other things too, so I don't think I've specialized enough in physics.

Thus, when Karin says that she is not specialised enough to be a physicist this is simultaneously a disidentification with a community she perceives to be too narrow for her in a negative, limiting sense and a disidentification with a community that is not available to her as an interdisciplinary researcher, someone who is ambitious rather than smart, who is a woman etcetera.

In summary, Karin could in many ways be described as being on the boundaries of physics; in terms of educational background, area of research and also gender. In her identity constitution the multidimensionality of the boundaries of the physicist community become apparent – and in all dimensions it seems as if she perceives the boundaries to be set in such a way that she is excluded. Thus, when Karin positions herself as a 'non-physicist' this is a highly complex positioning, taking place, as stated, against the backdrop of several perceived boundaries. Finally, it can also be noted that for Karin, as for Hanna, the lack of identification with the physicist community of practice is intertwined with a perceived inability to negotiate what it means to be a physicist; rather than trying to redefine – at least for herself – the boundaries of the physicist community of practice, the negotiation takes place on the level of her own identity.

9.3.4 Ann: A Physics Amateur

When I interviewed Ann as an undergraduate the relationship between gender and physics was very much in focus. In particular she talked about herself as a non-participant in a 'traditional' femininity. This non-participation is also present in the second interview as she, as a graduate student, describes not wanting to work in environments dominated by women. But even more pronounced in the second interview is a characterisation of herself and the other women at her department as 'laddish' – very similar to how Cecilia talks about herself. Like Cecilia, Ann thus both constitutes physics as gendered and her own participation in it as gendered. And, for both Cecilia and Ann their participation in physics makes a desired participation in a masculinity possible.

As an undergraduate student Ann strongly positioned herself as associating with the Discourse model of the analytical physics student. Now, as a PhD student, she says that this has changed markedly; drawing on her previous work experiences she emphasises her practical skills in the laboratory. This new positioning is one that she says became available during her Master's research project, where her supervisor was impressed by her practical skills. How Ann seems to find it easier to identify with the 'physicist', rather than the 'physics student', is also likely to be connected to age. Ann talks about how being at least ten years older than most of her course-mates made it difficult for her to fit into the undergraduate student community:

> Ann: I didn't want to barge in, and be the pathetic auntie, they were a group that studied together and then there was some odd guy who took one course with us, and then I did the labs with him.

Currently, as a graduate student, Ann is part of a research community were her age is no longer the anomaly she experienced it to be as an undergraduate. Norms of age are, in this case, thus tied to the general undergraduate student community, rather than the physicist community.

As a mature student Ann has reflected a lot on her choice to study physics and her participation in the physicist community of practice. Currently she is positioning herself as a physicist, a positioning that finally became available to her by the receiving of the degree certificate. In a sense the boundaries of the physicist community of practice are thus associated with having a university degree in physics. However, for Ann the boundaries of physics are much 'looser' than for Karin. Already in the first interview Ann talked about physics as very 'open'; how very different kinds of people could find a place in physics. Like for Karin, the physicist is for Ann, also associated with 'smartness', but their negotiation of their own participation in relation to this 'smartness' is very different. Whereas Karin finds it difficult to position herself as a physicist, finding the physicist community of practice to be too narrow, Ann re-defines the physicist community rather than re-defining her own identity as Karin does. For Ann there is no doubt about her positioning herself as a physicist, but she talks about herself as a 'physicist amateur', a leg-worker of physics. The positioning of herself as a leg-worker of physics can be interpreted as a 'doing of class'; Ann has a working class background and a positioning as a leg-worker creates a continuity from her participation in a workshop community to her current participation in the physicist community. Important to notice, however, is that Ann's positioning of herself as a leg-worker is not necessarily a degrading of her abilities; it can rather be interpreted as a negotiation of the physicist community of practice. The 'real work', she says, is done by ordinary people like herself, the 'extremely smart' are often just difficult to deal with. Consequently, the leg-workers of physics are, by Ann, constituted as central to the physicist community of practice, making her own positioning as a physicist a confident one.

9.4 Concluding remarks

In this chapter I returned to the students from Chapter 7 for a theoretically deepened exploration of their constitutions of identities as physicists and their constitutions of the practice of physics. The analysis in this chapter is founded in the student narratives from Chapter 7, but develops these further by drawing on the outcomes of Chapter 8 as well as insights provided by the application of the theoretical framework. In a sense this chapter thus ties the analysis together, with the aim of allowing my conceptual framework to bring the full complexity of the students' constitutions of physicist identities to the fore.

Next I will move on to the fourth, and final, part of this thesis, in which the theoretical and empirical outcomes are brought together, discussed, and used as the foundation for suggestions about future research. I will also discuss my study in relation to the teaching of physics.

PART IV

Discussion

CHAPTER 10

Concluding Discussion

The purpose of this thesis has been to, theoretically and empirically, explore how university physics students simultaneously constitute the practice of physics as enacted in the laboratory and constitute their own identities as physicists in that context. In particular, the thesis has focused on how these constitutions can be understood as gendered. In the following I have chosen to situate this concluding discussion in a number of **themes** that have emerged as being significant during my research process. I will also indicate what I feel would be interesting expansions of my work; ideas about future research. Finally, I discuss ways in which I believe my work can serve as an inspiration for the teaching and learning of physics in higher education.

For the theoretical purpose of the thesis I started with a broad framing of situated learning theory, which I complemented with post-structural gender theory as an integral part of the conceptual framework. To this I added analytical tools drawn from positioning theory and Gee's Discourse analysis, which completed my conceptual framework (Part II of the thesis). Reflecting on the research process I believe that this construction of the **conceptual framework** is a research outcome in itself, just as important as my analytical outcomes. Consequently, the conceptual framework became both a tool for my research and an outcome of my research. However, despite having given my conceptual framework the status of a research outcome, I fully appreciate that it can be strengthened further and this is one of my immediate future aims. Some of these possible developments will be outlined in the discussion that follows.

As a physicist, a physics education researcher, it has been of crucial importance for me to empirically and theoretically explore the students' **doing of physics** (and not science in general). In Chapter 2 I argued that it has been common in much previous research to approach the science disciplines as a unity; a single science community. While I acknowledge the importance of this research I would also argue that it needs to be complemented with studies that take the particularities of the different scientific disciplines into account. For example, when women in my empirical material are constructed as an 'anomaly' in relation to the practice, this is not done in relation to a general scientific practice, or even a general physics practice – but in many cases only an anomaly in relation to the practical aspects of laboratory work in physics. I will expound on this below, but first a few words about the Discourse models.

In my exploration of the practice of the student laboratory I constructed two physics student **Discourse models** (the Discourse model of the analytical physics student and the Discourse model of the practical physics student). In important ways these Discourse models reflect different facets of the 'doing of physics', and at the same time they are tightly intertwined with the students' 'doing of gender' and 'doing of class'. For example, both the physics student Discourse models can be regarded as connected to the technological, partly-class based, masculinities described by Wajcman (1991). However, the Discourse model of the analytical physics student also overlaps with what could be characterised as a traditional science-student femininity (the portrayal of woman science students as diligent, neat, thorough and rule-following). The characterisation of such a femininity also brings to the fore how the masculine norms of physics are not the only gender norms woman physics students may have to relate to; there are also norms for how to be a woman physics student. Thus, not only are the woman physics students relating to masculine norms of the discipline, they may also have to deal with the norms and expectations about how a woman is supposed to be in a physics and engineering context. Several of the woman students in the study did negotiate a potentially disempowering participation in such a traditionally science student femininity (taking on the secretarial duties in the laboratory is not likely to give one status) by tying it to aspects of the high status Discourse model of the analytical physics student. They thereby could constitute a strong physicist identity without having to distance themselves from the traditional science student femininity. A somewhat unexpected negotiation of these norms was also exemplified by 'Mia', who was able to constitute a comfortable physicist identity by drawing on a discourse of women as having a natural aptitude for logic and mathematics. However, there is also the possibility of a positioning that strongly challenges the norms for femininity within science; among the students interviewed 'Cecilia' was probably the woman with the most outspoken resistance towards these norms. Not only was she resisting the traditional science student femininity, by describing an unstructured and unprepared approach to the student laboratory, she was also resisting femininity in more general terms and repeatedly referred to herself as 'laddish'. She also pointed out how this 'laddishness' contributed to it being easier for her than other women to fit within the boundaries of physics. Such a distancing from a traditional femininity among women within science and technology as a way to fit into a man-dominated discipline is well documented in previous research. However, a counter-identification with traditional femininity is often interpreted in negative terms, either as a failure to challenge dominant gender norms (Henwood

1996, 1998) or even as an approach forced onto the women (Kvande 1999). While I do not oppose these interpretations I would like to use the case of 'Cecilia' to give nuance to the picture: What happens if we interpret 'Cecilia's' choice of physics and her approach to laboratory work in terms of doing of a female masculinity (Halberstam 1998)? If we understand 'Cecilia's' participation in physics as enabling a simultaneous participation in a desired female masculinity? These are questions that may provide a nouveau entrance point to the exploration of women's constitutions of identities as physicists. From this perspective 'Cecilia's' counter identification with traditional femininity may not challenge the masculine norms of physics, but it does challenge heteronormative understandings of all men as desiring to be masculine and all women desiring to be feminine. Furthermore, 'Cecilia's' unafraid and unplanned tinkering in the laboratory also challenge disciplinary gender norms, the traditional science student femininity.

So far I have discussed how the Discourse models can be understood as relating to the students' doing of gender. Yet, one of the most prominent themes in the empirical material in regard to the students' constitutions of gender is in fact the constitution of **gender neutrality**, both in terms of physics as a discipline and in terms of academia in general. This is done in part by stressing of individual characteristics and abilities as mattering, disregarding the impact of structural forces. Such an attribution of academic achievement purely to individual skills and motivation was also identified by Erwin and Maurotto (1998) in their study of woman undergraduate science students; even when the students did acknowledge structural forces these were constructed as obstacles that could be overcome, given enough effort and motivation. Erwin and Maurutto further concluded that a consequence of this individualistic discourse was that 'few [of the interviewed women] appeared to know strategies, other than those involving self-improvement, that would help them cope with the barriers and problems they confronted' (Erwin and Maurotto 1998, p. 60). And, I would like to add, it makes the appropriate participation in physics a gender neutral one. Another way the gender neutrality is constructed is through the contrasting of academia or physics to other, presumed non-gender neutral milieus, such as in 'Klara's' description below:

I: So even if you have equality in your group, physics as a whole is quite dominated be men, how do you experience being a woman in the physics environment?

Klara: I haven't experienced any problems or anything during the education really. But I have thought about it before, that I haven't experienced it. So... and not now during the Master's research project either, but... We were on a conference some time ago and then... a CEO I think it was had an opening speech and then... he showed a picture of underwear, about positive effects of global warning and then he started with a really big pair and ended up with gee-strings and then all the old men laughed and then I felt, 'but, oh, it's not

> funny'. And then I realised, it's not so easy all the time [laughter] to be a woman kind of. And we weren't a lot of women at that conference, it was just...

So, despite the dominance of men in physics Klara is able to construct physics as gender neutral by using the corporate world as contrast, thereby also making her earlier claim about the equality of her group a believable one. In total, the construction of physics as gender neutral is a complex construction that needs to be understood against the backdrop of both disciplinary ideals of objectivity as well as academic discourses about individuality and meritocraty and a Swedish discourse of gender equality. But, the construction of physics (and academia in general) as gender neutral can also be tied to a fear that naming problems, as related to gender, may confirm beliefs about women as incapable of succeeding in physics (Henwood 1998).

An important focus of my study has been 'gender', in particular in the construction of my theoretical framework. However, in the empirical investigation the need to deal with the **intersections** between gender, age and class have become increasingly apparent – and how the analytical tool of positioning would also allow for the analysis of such intersections. Furthermore, intersectionality (Lykke 2003) is one of the areas were a further development of the conceptual framework would be interesting. Such a development would, for example, contribute to addressing issues of how class and gender are simultaneously done in the discipline of physics. The development of my conceptual framework was founded in an ambition to find a way to simultaneously explore students' doing of physics and doing of gender – a reaction towards how previous research has often had a disproportionate focus on either the doing of gender or the doing of science – and thus, the inclusion of other aspects such as class or age would be a natural continuation.

In a study employing a communities of practice perspective an obvious question to ask is how the abstract entity of the 'physicist community of practice' can be defined and delimitated. I began the discussion of this in section 8.5 but will here share a few more thoughts on the issue. One way to approach the issue could be to create a point of departure from small, localised communities (see, for example, Solomon 2007). In my case such communities could be classes of physics students; and then explore what characterises these, how such communities relate to one another; to look for commonalities and differences. However, in agreement with Nespor (1994) I believe that such a focus on small-scale communities is far too insular in the case of physics. It presupposes boundaries where there may be none and disregards the importance of wider disciplinary networks. So rather than focusing my attention on how the physicist community ought to be defined, I have started from a rather flexible understanding of this community and focused my attention on the **boundary work** done by the students: How are they constituting the boundaries relevant to their participation in physics? In what ways can these boundaries be excluding and including? How are they negotiating their own and others participation in physics? This flexible understanding of the physicist community of practice has allowed me to build my analysis of the constitution of the community from the empirical material, but also made use of the theoretical strengths offered by a communities of practice perspective.

But while the students interviewed in this study most certainly were constructing and reconstructing various boundaries in and around the physicist community of practice, there was also a tendency among them to describe an absence of temporal boundaries in their relation to science: A scientist is nothing they have become; it is something they have always been. This narration of the scientist as a stable, authentic part of oneself is a well-established one within the science communities and can as such be understood as a part of the shared repertoire of such communities. As an example, consider the following quote from the well-know physicist Richard Feynman's (1992) essay 'The Amateur Scientist':

> When I was a kid I had a 'lab'. It wasn't a laboratory in the sense that I would measure or do important experiments. Instead I would play: I'd make a motor, I'd make a gadget that would go off when something passed a photocell. I'd play around with selenium; I was piddling around all the time.
>
> (Feynman 1992, p. 91)

Similar stories are also found in my interview transcripts; of childhood fascinations for astronomy or the wish to hear popular science books as bedtime stories. But perhaps this narration of childhood interests in science is particularly important for the woman students; a way for them to claim **authenticity** in their constitutions of physicist identities. In order to do so their stories may also stress how their childhood was different from that of other children, other girls. Marina, for example, not only talks about always having had technical interests, by saying that these interests were not 'a problem' to her parents, she also acknowledges the unexpectedness of them:

> I: But what was it that attracted you to physics, apart from it being difficult?
>
> Marina: Erm... Yeah, well, it was very much that I wanted to prove myself, mostly to myself really. And my parents have no problems with me having such technical interests. And... yeah, I've always wanted to know how things work, how they are connected and... Basically, to understand the processes better.

Thus, constituting an identity as a physicist is not something done only in the present, it can also involve telling the right kind of story about a temporarily stable interest in the discipline. The importance of this stable interest, of the childhood experiments for the professional physicist, is further reinforced in stories such as the following told by the Nobel laureate in physics Carl E.

Wieman, who describes his talent for experimental physics as founded in childhood constructions and investigations (notice the similarity to the Discourse model of the practical physics student):

> Brook and I also spent many hours engrossed in all sorts of projects constructing and investigating things. I think that much of my talent and enjoyment at improvising solutions to experimental problems goes back to those homebuilt projects. In this regard my older brother Howard also inspired me; he was always tinkering with machines and building astonishingly elaborate toys for his younger siblings. Carrying out these individual projects also developed in me a good sense of self-reliance and a sense when a piece of improvised apparatus was likely (or unlikely) to be adequate. This sense is one that I often see missing in students whose education has been confined to formal instruction.
>
> (Wieman 2002)

In this story it is also noticeable how Wieman (2002) stresses how formal instruction is not enough to develop the skills necessary for the physicist. This theme of how the participation in the physicist community of practice in part is made legitimate through a correct narration of the 'past self' is one that deserves more attention and one that would be interesting to explore further in future research.

Finally, there is another theme that deserves further attention – **power relations**. In this thesis I have discussed this in relation to situated learning theory and how several researchers have brought the lack of attention to power relations forward as a weakness in situated learning theory. Reflecting on my own study I realise that I have not been able to be as mindful of issues of power when doing my analysis as I would have wanted. Mindfulness in regard to issues of power becomes perhaps especially pertinent in working with a high status discipline like physics. It is not uncommon for physicists to rank their discipline as the one with the highest status (see, for example, Benckert and Staberg 2000) and coupled with this status are also conceptions of the intelligence needed to work in the discipline (Traweek 1988). As pointed out by Henwood (1998), women who choose to study physics not only make a different choice than is expected, they have chosen a discipline with higher status; that is associated with men, masculinity and masculine power. The power associated with the discipline can most certainly be attractive, like for Marina, who explains that she chose Engineering Physics because it is the most difficult education you can study:

I: Yes, you're studying Engineering Physics, yeah... Why did you chose that education?

Marina: Erm... 'Cause I never got any boundaries in school and I talked to people and wanted to know what the most difficult education was. And then people agreed that Engineering Physics is definitely the toughest thing you can study.

This is not unlike Evelyn Fox Keller (1977), who I quoted in my introduction, who describes being attracted to physics because of the image of herself 'striving and succeeding in an area where women had rarely ventured' (Fox Keller 1977, p. 78). However, in discussions about women and physics the other side of the story is more commonly told, how the constitution of physics as difficult, high status discipline can be problematic for women – like, for example, 'Karin', for whom it was very important to explain that she chose an engineering education because of her interests, not the status associated with the education. This brief empirical example demonstrates how the high status of a discipline can be both attractive and problematic and more detailed investigations of the diverse workings of power relations in and across communities of practice is called for in future research.

Next I will share some additional reflections on my study, before the thesis is rounded off with a section which I anticipate will serve as inspiration for teachers.

10.1 Reflections on the Research Process

This thesis is the end-product of a study of the learning of physics that began in the spring of 2004. Coming to the end of my study allows me to reflect on the research process; decisions made and routes taken – in particular those that perhaps could have been done differently. For example, early on in the process I decided to work with interviews; it was important for me to let the students voice their own experiences, rather than interpreting their activities in the laboratory from the view-point of an observer. Today I can see the advantages of involving the interviewees more in the actual research process; how, for example, letting the interviewee reflect on an interview transcript and/or the emerging analytical outcomes could add depth and richness to the analysis. This would also make it possible to capture development and change to a greater extent. However, at the time of the interviews a deeper involvement of the interviewees in the research process would probably not have been possible. I was learning the handicraft of research and I feel that involving the interviewees further would have added too much complexity. A deeper involvement with the interviewees could also have included participant observations, in particular joining the Master's and PhD students in their daily work would have added increased depth to those interviews.

During the analytical process I have been working both in English and Swedish – Swedish transcripts with English translations of the interview excerpts included in the thesis – something which has added an additional challenge to the analysis and especially the presentation of it in this thesis. Most certainly nuances get lost in the translation and it is difficult to give justice to the interviewees' expressions when translated.

10.2 Inspiration for teaching

This is a thesis about the learning of physics; how students learn to become physicists. It explores how students, in this becoming, simultaneously with their doing of physics can also be understood as doing gender. Throughout the thesis 'learning' is understood not only as the result of formal instruction but also as something involving broader personal and societal contexts. In short, learning as an identity constitution. However, this does not make it any less interesting to reflect upon the outcomes of my study in the light of how they can inspire our teaching of physics – and that is what this last section of the thesis will focus on.

The first thing I would like to point out in regard to the issue of my thesis as an inspiration for teaching is that I believe that the most important 'pedagogical implications' of my research are not the ones that I can explicitly provide in this section – they are the ones that have already taken place in the meeting between my text and the (physics) teacher reading it. In other words, I, like Clandinin (1992), hope that my thesis will 'have readers raise questions about their practice, their ways of knowing' (Clandinin 1992, p. 135). In the broadest sense, my ambition with this thesis, from a teaching point of view, is to inspire physics teachers to reflect on their practice from novel perspectives.

On a more detailed level, the various parts of the thesis may also serve as 'thinking devices' for our teaching: The Discourse models may make us more aware of students' different approaches in the laboratory. The students' positionings of themselves and others (including the teacher) may make us reflect upon how those positionings create inclusion, exclusion and hierarchies in the classroom. The student narratives may open our eyes to how gender is done in the physics classroom.

An important part of these reflections on our own teaching practice is getting to know our students; what they require and desire in the teaching situation. In this respect, our own personal experiences may, as pointed out by Redish (2003), be a very poor guide for telling us the best way to teach our students. He writes:

> …I can't remember a time when I couldn't read a graph, and I have found it difficult to empathize with students who come to physics and can't read a graph or reason proportionately. It takes a special effort for me to figure out the right approach.
>
> (Redish 2003, p. 38)

Thus, what Redish is saying is that we cannot expect the students in our classroom to be junior versions of ourselves – and physics education research can provide important insights into students' diverse ways of experiences the learning of physics. So, on one level this thesis may provide us as physics teachers with examples of students we may meet in the classroom, thereby enhancing our understanding of possible learning experiences. But, importantly, the students' stories in this thesis are stories that have been made sense of through theories about learning and about gender, which have contributed to insights beyond those of the original story-tellers, the students. On one hand, such theoretical explorations do of course have value in their own right, like any other kind of basic research. However, they can also be illuminating and empowering for the individual student on very concrete terms; giving them tools for reflecting on their own experiences and impressions. This is especially pertinent since both my study and earlier research in the field show that many science students do not have strategies for making sense of their participation in science beyond an individualistic discourse that attributes academic achievement purely to individual skills and motivation (see, for example, Erwin and Maurutto 1998; Lundborg and Schönning 2007). From this perspective, it is my hope that my thesis, in particular the student narratives in Chapter 9 with their heightened level of theoretical abstraction, can provide physics students with tools for reflecting upon their own participation in physics.

But how about physics teachers reflections about their own, and their students, participation in physics? This is something I would like to discuss in the light of the notion of scientific literacy. Scientific literacy is often brought forward as a central goal in science education, in particular at school level (see, for example, Linder et al. 2007). Roberts (2007a, 2007b) has characterised two schools of thought within discussion about what kind of literacy school science ought to promote, Vision I and Vision II. According to Vision I, school science should aim to enable students to 'approach and think about situations' in a way similar to professional scientists; to be literate in science. In Vision II the aim is to educate students to be able to 'approach and think about situations' as informed citizens; to be literate about science, have knowledge about socio-scientific issues (Roberts, 2007). I would like to propose that we take Roberts' notion of Vision II to reflect on *our* own scientific literacy in a way that includes what sometimes is referred to as cultural literacy. In short, cultural literacy of science involves an understanding of science as a social process; how 'nature-as-an-object-of-knowledge is always cultural' (Harding 1993, p. 1). It involves what could be characterised as a critical outsider's perspective of science, the ability to critically examine science in terms of, for example, its strengths and weaknesses, knowledge claims and limitations. This is something that is not typically included in science educations, which:

> ...rarely expose students to systematic analyses of the social origins, traditions, meanings, practices, institutions, technologies, users, and consequences of the natural sciences that ensure the fully historical character of the results of scientific research.
>
> (Harding 1993, p. 1)

A teaching of physics engaging issues of cultural literacy would, thus, most fundamentally require us not only to teach physics, but also to teach about physics. The advocating of such a teaching is, of course, nothing new, for example, Gilbert (2001), Letts (2001) and Kumashiro (2001) have argued for a science education that gives students tools to critically examine scientific knowledges and practices. However, their argument is primarily concerned with science at school level, Gilbert (2001) writes:

> ...science education programmes should aim to educate students *about* science, not to train them to *be* scientists. Its aim is to develop students' critical 'literacy' with respect to science, to develop a relationship to science which is similar to the one that a literary critic or art critic might have to literature or art: that is, one which is rather different from *being* a writer or an artist.
>
> (Gilbert 2001, p. 300, italics in original)

Following Harding (1993) my argument is that the promotion of the type of critical examination of science advocated by Gilbert (2001) also has its place in programmes educating students to become scientists. My ambition is that this thesis can promote discussion about gender and physics (education) by inviting readers, students and teachers alike, to engage in critical reflections about the practice of physics. In short, to allow them to further develop their cultural literacy of physics.

Acknowledgments

During my time as a PhD student I have had the privilege to work in two very different but equally stimulating and creative research environments: The division of physics education research at the Department of Physics and Materials Science and the Centre for Gender Research. All my colleagues at both these research environments are gratefully acknowledged for inspiring discussions and endless support. This thesis would not have been possible without you. Further, I would like to express my sincere gratitude to the following people:

My supervisor Cedric Linder. For introducing me to physics education research and for, with extensive knowledge and everlasting encouragement, guiding me through my entire PhD education. For giving me the possibility to pursue my own research interests and for giving a much valued independence that has allowed me to grow and develop as a researcher.

My three co-supervisors: Leif Karlsson, Rebecca Lippmann Kung and Tora Holmberg. They have all in their own ways contributed to making this thesis what it is today. Leif Karlsson, for being my guide in the world of atomic and molecular physics and arranging my research visit the University of Oxford (where John Eland so kindly took me into his research group). Rebecca Lippmann Kung, for thoughtful comments on my work in the introductory phases, for giving so freely of your time and for introducing me to new perspectives on physics education research. Tora Holmberg, for looking at my work from new perspectives, challenging and inspiring me, and helping me to develop the gender perspective of the thesis. In this context I would also like to mention Melanie Walker whose inspiring supervision regarding issues of gender and education early on helped me to develop those aspects of my work.

Past and present members of the Physics Education Research Group: Daniel Domert, Liselott Dominicus, Anne Linder, John Airey, Johan Falk, Bjørn Friis Johannsen, Staffan Andersson and Jonas Forsman. For intellectually challenging discussions as well as their general support and many enjoyable moments. A special thanks to Anne Linder for her proficient language check of this thesis.

At the Centre of Gender Research I would like, in particular, to mention my colleagues within the research profiles *Physics* and *Education*. The warmest thanks also to Staffan Wennerholm, Lisa Folkmarson Käll and Kristina Eriksson, who at various stages of the research process have shared their insightful comments on my text. And, of course, my PhD student colleagues: Henrik Berg (also the perfect room-mate!), Anneli Andersson, Elin Bjarnegård and Maria Margareta Österholm.

Twice during my PhD period I have had the privilege to have my work thoroughly examined by external readers, at my licentiate seminar and at my final seminar. For this I want to thank Minna Salminen-Karlsson and Anne-Sofie Kalat, as well as all the participants in the final seminar. I am also very grateful to Ann-Charlotte Ek, who was the first outside-colleague to read a full thesis draft and give insightful feedback and encouraging comments.

Åke Ingerman and his colleagues in the PLinG (Physics Learning in Groups) project, for inviting me to participate in your research project. This also allowed me to look at my thesis work from novel perspectives.

Slutligen – och allra viktigast – min familj och mina vänner: Tack för allt!

CHAPTER 11

References

Aberg-Bengtsson, L., and Ottosson, T. (2006). What lies behind graphicacy? Relating students' results on a test of graphically represented quantitative information to formal academic achievement. *Journal of Research in Science Teaching, 43*(1), 43-62.

Adams, W. K., Perkins, K. K., Podelfsky, N. S., Dubson, M., Finkelstein, N. D., and Wieman, C. E. (2006). New instrument for measuring student beliefs about physics and learning physics: The Colorado Learning Attitudes about Science Survey. *Physical Review Special Topics - Physics Education Research, 2*, 1-14.

Airey, J., and Linder, C. (2009). A Disciplinary Discourse Perspective on University Science Learning: Achieving Fluency in a Critical Constellation of Modes. *Journal of Research in Science Teaching, 46*(1), 27-49.

Alexopoulou, E., and Driver, R. (1997). Gender differences in small group discussion in physics. *International Journal of Science Education, 19*(4), 393-406.

Allie, S., Buffler, A., Campbell, B., Lubben, F., Evangelinos, D., Psillos, D., and Valassiades, O. (2003). Teaching Measurement in the Introductory Physics Laboratory. *The physics teacher, 41*, 394-401.

Allie, S., Buffler, A., Kaunda, L., and Inglis, M. (1997). Writing-Intensive Physics Laboratory Reports: Tasks and Assessment. *The physics teacher, 35*, 399-340.

Ambjörnsson, F. (2004). *I en klass för sig. Genus, klass och sexualitet bland gymnasietjejer.* [In a special class. Gender, class, and sexuality among girls in high school] Stockholm: Ordfront Förlag.

Ambjörnsson, F. (2006). *Vad är queer?* [What is queer?] Stockholm: Natur och kultur.

Ambrose, B. S., Shaffer, P. S., Steinberg, R. N., and McDermott, L. C. (1999). An investigation of student understanding of single-slit diffraction and double-slit interference. *American Journal of Physics, 67*(2), 146-155.

Ambrosia, S. M. (1940). Teaching Physics to Women. *American Journal of Physics, 8*(5), 289-290.

Angell, C., Guttersrud, Ö., and Henriksen, E. K. (2004). Physics: Frightful, but fun. Pupils' and teachers' views of physics and physics teaching. *Science Education, 88*, 683-706.

Barnes, M. (2004). *The use of positioning theory in studying student participation in collaborative learning activities.* Paper presented at the Annual Meeting of the Australian Association for Research in Education, Melbourne.

Bazley, M., Herklots, L., and Branson, L. (2002). Using the Internet to make physics connect. *Physics Education, 37*(2), 118-121.

Benckert, S. (1997). Är fysiken könlös? Reflektioner kring ett universitetsämne. [Has physics no gender? Reflections about a university discipline] In G. Nordborg (Ed.), *Makt & kön* [Power & gender] Eslöv: B. Östlings Symposion.

Benckert, S. (2005). Varför väljer inte flickorna fysik? [Why do not girls choose physics?] *Tidskrift för lärarutbildning och forskning* [Journal of Research in Teacher Education] *12*(4), 27-37.

Benckert, S., and Pettersson, S. (2008). Learning Physics in Small-Group Discussions - Three Examples. *Eurasia Journal of Mathematics, Science & Technology Education, 4*(2), 121-134.

Benckert, S., and Staberg, E.-M. (2000). *Val, villkor, värderingar: Samtal med kvinnliga fysiker och kemister.* [Choices, conditions, and values. Conversations with female physicists and chemists] Umeå Universitet. Kvinnovetenskapligt forum.

Berge, M., and Danielsson, A. (2008). *The storyline of learning physics in groups - perspectives on becoming a physicist.* Paper presented at the Higher Education CLose-up 4 Conference, Cape Town, South Africa.

Berner, B. (2003). Kön, teknik och naturvetenskap i skolan. [Gender, technology, and science in school] In B. Berner (Ed.), *Vem tillhör tekniken? Kunskap och kön i teknikens värld* [Who does technology belong to? Knowledge and gender in the world of technology] (pp. 119-144). Lund: Arkiv förlag.

Berner, B. (2004). *Ifrågasättanden. Forskning om genus, teknik och naturvetenskap* [Questionings. Research about gender, technology and science] Tema T, rapport 42. Linköping.

Blumenfeld-Jones, D. (1995). Fidelity as a criterion for practicing and evaluating narrative inquiry. In J. A. Hatch and R. Wisniewski (Eds.), *Life history and narrative* (pp. 23-35). London: Falmer Press.

Bogdan, R. C., and Biklen, S. K. (1996). *Qualitative Research for Education: An introduction to theory and methods.* Boston: Allyn & Bacon.

Brickhouse, N. W. (2001). Embodying Science: A feminist perspective on learning. *Journal of research in science teaching, 38*(3), 282-295.

Brickhouse, N. W., Lowery, P., and Scultz, K. (2000). What kind of girl does science? The construction of school science identities. *Journal of research in science teaching, 37*(5), 441-458.

Brickhouse, N. W., and Potter, J. T. (2001). Young Women's Scientific Identity Formation in an Urban Context. *Journal of research in science teaching, 38*, 965-980.

Brookes, D. T., and Etkina, E. (2007). Using conceptual metaphor and functional grammar to explore how language used in physics affects student learning. *Physical Review Special Topics - Physics Education Research, 3*, 1-16.

Brown, J. S., and Duguid, P. (1991). Organizational Learning and Communities-of-Practice: Toward a Unified View of Working, Learning, and Innovation. *Organization Science, 2*(1), 40-57.

Buffler, A., Allie, S., Lubben, F., and Campbell, B. (2001). The development of first year physics students' ideas about measurement in terms of point and set paradigms. *International Journal of Science Education, 23*(11), 1137-1156.

Butler, J. (1999). *Gender Trouble. Feminism and the subversion of identity.* New York: Routledge.

Byrne, E. (1993). *Women and science: The snark syndrome.* London: Falmer Press.

Calabrese Barton, A. (1997). Liberatory science education: Weaving connections between feminist theory and science education. *Curriculum Inquiry, 27*(2), 141-163.

Carlone, H. B. (2004). The cultural production of science in reform-based physics: girls' access, participation, and resistance. *Journal of Research in Science Teaching, 41*(4), 392-414.

Case, J., and Jawitz, J. (2004). Using Situated Cognition Theory in Researching Student Experience of the Workplace. *Journal of research in science teaching, 41*(5), 415-431.

Case, J., and Marshall, D. (2006). The 'no problem' Discourse model: Exploring an alternative way of researching student learning. *Unpublished manuscript.*

Chu, H.-E., Treagust, D. F., and Chandrasegaran, A. L. (2008). Naïve Students' Conceptual Development and Beliefs: The Need for Multiple Analyses to Determine what Contributes to Student Success in a University Introductory Physics Course. *Research in Science Education, 38*(1), 111-125.

Clandinin, D. J. (1992). Narrative and story in teacher education. In T. Russell and H. Munby (Eds.), *Teachers and teaching: from classroom to reflection* (pp. 124-137). London: Falmer Press.

Clement, J. (1982). Students' preconceptions in introductory mechanics. *American Journal of Physics, 50*(1), 66-71.

Cohen, L., Lawrence, M., and Morrison, K. (2000). *Research Methods in Education* (5th ed.). London: Routledge Falmer.

Colin, P., and Viennot, L. (2001). Using two models in optics: Students' difficulties and suggestions for teaching. *Physics Education Resources, American Journal of Physics, Suppl., 69*(7), S36-S44.

Connell, R. W. (2003). *Maskuliniteter.* [Masculinities] Göteborg: Daidalos.

Cox, A. J., and Junkin, W. F. (2002). Enhanced student learning in the introductory physics laboratory. *Physics Education, 37*(1), 37-44.

Crouch, C. H., Fagen, A. P., Callan, J. P., and Mazur, E. (2004). Classroom demonstrations: Learning tools or entertainment? *American Journal of Physics, 72*(6), 835-838.

Crouch, C. H., and Mazur, E. (2001). Peer Instruction: Ten years of experience and results. *American Journal of Physics, 69*(9), 970-977.

Danielsson, A. T. (2007). *The gendered doing of physics: A conceptual framework and its application for exploring undergraduate physics students' identity formation in relation to laboratory work.* Unpublished Licentiat thesis, Uppsala University, Uppsala.

Danielsson, A. T. (forthcoming). Gender in physics education research: A review and a look forward. In M. Blomqvist and E. Lindsmyr (Eds.), *Never mind the gap! Gendering Science in Transgressive Encounters* [working title]. Centrum för genusvetenskap: Uppsala.

Danielsson, A. T., and Linder, C. (2009). Learning in Physics by doing Laboratory Work: towards a new Conceptual Framework. *Gender and Education, 21*(2), 129-144.

Davidowitz, B., and Rollnick, M. (2003). Enabling Metacognition in the Laboratory: A Case Study of Four Second Year University Chemistry Students. *Research in Science Education, 33*, 43-69.

Davies, B. (2005). Community of practice: legitimacy not choice. *Journal of Sociolinguistics, 9*(4), 557-581.

Davies, B. (2006). Subjectification: the relevance of Butler's analysis for education. *British Journal of Sociology of Education, 27*(4), 425-438.

Davies, B., and Gannon, S. (2005). Feminism/Poststructuralism. In B. Somekh and C. Lewin (Eds.), *Research Methods in the Social Sciences: A Guide for Students and Researchers.* (pp. 318-325) London: SAGE.

Davies, B., and Harré, R. (1999). Positioning and Personhood. In R. Harré and L. van Langenhove (Eds.), *Positioning Theory: Moral Contexts of Intentional Action.* (pp. 32-52) Oxford: Blackwell Publications.

Ding, N., and Harskamp, E. (2006). How partner gender influence female students' problem solving in physics education. *Journal of Research in Science Education and Technology, 15*(5), 331-343.

diSessa, A. (1993). Toward an Epistemology of Physics. *Cognition and Instruction, 10*(2&3), 105-226.

Domert, D., Airey, J., Linder, C., and Lippman Kung, R. (2007). An exploration of university physics studentss epistemological mindsets towards the understanding of physics equations. *Nordic Studies in Science Education, 3*(1), 15-28.

Domert, D., Linder, C., and Ingerman, Å. (2005). Probability as a conceptual hurdle to understanding one-dimensional quantum scattering and tunnelling. *European Journal of Physics, 26*, 47-59.

Du, X.-Y. (2006). Gendered practices of constructing an engineering identity in a problem-based learning environment. *European Journal of Engineering Education, 31*(1), 35-42.

Duit, R., Häussler, P., Lauterbach, R., Mikelskis, H., and Westphal, W. (1992). Combining issues of "girl-suited" science teaching, STS and constructivism in a physics textbook. *Research in Science Education, 22*, 106-113.

Duit, R., and Treagust, D. F. (2003). Conceptual change: a powerful framework for improving science teaching and learning. *International Journal of Science Education, 25*(6), 671-688.

Eduards, M. (2007). *Kroppspolitik. Om Moder Svea och andra kvinnor.* [Bodily politics. About Moder Svea and other women] Stockholm: Atlas Akademi.

Elby, A., and Hammer, D. (2001). On the substance of a sophisticated epistemology. *Science Education, 85*(5), 554-567.

Elvin-Nowak, Y., and Thomsson, H. (2003). *Att göra kön. Om vårt våldsamma behov av att vara kvinnor och män.* [Doing gender. About our urgent need to be women and men] Stockholm: Albert Bonniers Förlag AB.

Ely, M. (1991). *Doing qualitative research: circles within circles*. London: Falmer Press.

Erwin, L., and Maurutto, P. (1998). Beyond access: considering gender deficits in science education. *Gender and Education, 10*(1), 51-69.

Etkina, E., Gibbons, K., Holton, B. L., and Horton, G. K. (1999). Lessons learned: A case study of an integrated way of teaching introductory physics to at-risk students at Rutgers University. *American Journal of Physics, 67*(9).

Etkina, E., Warren, A., and Gentile, M. (2006). The Role of Models in Physics Instruction. *The physics teacher, 44*(1), 34-39.

Feynman, R. P. (1992). *Surely you're joking Mr. Feynman. Adventures of a curious character*. London: Vintage Books.

Finegold, M., and Gorsky, P. (1991). Students' concepts of force as applied ro related physical systems: A search for consistency. *International Journal of Science Education, 13*(1), 97-113.

Ford, D. J., Brickhouse, N. W., Lottero-Perdue, P., and Kittleson, J. (2006). Elementary Girls' Science Reading at Home and School. *Science Education, 90*(2), 270-288.

Forster, P. (2005). Alternative physics examination questions: identification and explanation of different discriminating powers. *Research in Science Education, 35*(4), 395-423.

Fox Keller, E. (1977). The anomaly of a woman in physics. In S. Ruddick and P. Daniels (Eds.), *Working It Out* (pp. 78-91). New York: Pantheon Books.

Fox Keller, E. (1992). *Secrets of Life. Secrets of Death.* London: Routledge.

Friis Johannsen, B. (2007). *Attrition in University Physics. A narrative study of individuals reacting to a collectivist environment.* Unpublished Licentiat Thesis, Uppsala University, Uppsala.

GASAT (Gender and Science and Technology) (2009) Retrieved February 23, 2009, from http://www.gasat-international.org

Gee, J. P. (2005). *An introduction to discourse analysis. Theory and method.* (2nd ed.). London: Routledge.

Georghiades, P. (2004). From the general to the situated: three decades of metacognition. *International Journal of Science Education, 26*(3), 365-383.

Gilbert, J. (2001). Science and its 'Other': looking underneath 'woman and 'science' for new directions in research on gender and science education. *Gender and Education, 13*(3), 291-305.

Gilbert, J., and Calvert, S. (2003). Challenging accepted wisdom: looking at the gender and science education question through a different lens. *International Journal of Science Education, 25*(7), 861-878.

Gill, J., Mills, J., Franzway, S., and Sharp, R. (2008). 'Oh you must be very clever!' High-achieving women, professional power and the ongoing negotiation of workplace identity. *Gender and Education, 20*(3), 223-236.

Gisselberg, K., Ottander, C., and Hanberger, A. (2004). *NOT-projektet 1999-2003 - en utvärdering.* [The NOT-project 1999-2003 – an evaluation] Umeå: Umeå University.

Glasser, H. M., and Smith III, J. P. (2008). On the Vague Meaning of "Gender" in Education Research: The Problem, Its Sources, and Recommendations for Practice. *Educational Researcher, 37*(6), 343-350.

Goldberg, F., and McDermott, L. C. (1987). An investigation of student understanding of the real image formed by a converging lens or concave mirror. *American Journal of Physics, 55*(2), 108-119.

Guba, E. G., and Lincoln, Y. S. (1989). *Fourth Generation Evaluation.* Newbury Park: Sage.

Gunstone, R. (1991). *Constructivism and metacognition: Theoretical issues and classroom studies.* Paper presented at the Research in Physics Learning: Theoretical Issues and Empirical Studies, University of Bremen.

Gustafsson, P. (2005). Gender inclusive physics education - a distance case. *European Journal of Physics, 26*, 843-849.

Hake, R. H. (1996). Interactive-engagement versus traditional methods: A six-thousand-student survey of mechanics test data for introductory physics courses. *American Journal of Physics, 66*, 64-74.

Halberstam, J. (1998). *Female masculinity.* Durham: Duke University Press.

Halloun, I. A., and Hestenes, D. (1985a). The initial knowledge state of college physics students. *American Journal of Physics, 53*(11), 1043-1048.

Halloun, I. A., and Hestenes, D. (1985b). Common sense conceptions about motion. *American Journal of Physics, 53*(11), 1056-1065.

Hammer, D. (1995). Epistemological considerations in teaching introductory physics. *Science Education, 79*(4), 393-413.

Hammer, D. (1996). More than misconceptions: Multiple perspectives on student knowledge and reasoning, and an appropriate role for education research. *American Journal of Physics, 64*(10), 1316-1325.

Hammer, D., and Elby, A. (2002). On the Form of a Personal Epistemology. In B. K. Hofer and P. R. Pintrich (Eds.), *Personal Epistemology: The Psychology of Beliefs about Knowledge and Knowing*. (pp. 169-190) Mahwah, NJ: Erlbaum.

Hammer, D., Elby, A., Scherr, R., and Redish, E. F. (2004). Resources, framing, and transfer. In J. Mestre (Ed.), *Transfer of Learning: Research and Perspectives*. (pp. 89-120) Greenwich, CT: Information Age Publishing.

Harding, J. (1979). Sex differences in examination at 16+. *Physics Education, 14*, 280-284.

Harding, S. (1986). *The science question in feminism*. Milton Keynes: Open university press.

Harding, S. (2005). Science and Technology. In P. Essed and D. T. Goldberg and A. Kobayashi (Eds.), *A Companion to Gender Studies*. (pp. 241-254) Malden, MA: Blackwell.

Harding, S. (1993). Introduction. In S. G. Harding (Ed.), *The "Racial" Economy of Science. Toward a Democratice Future.* (pp. 1-29) Bloomington: Indiana University Press.

Harding, S. (1998). *Is Science Multurcultural? Postcolonialisms, feminisms, and epistemologies*. Bloomington, IN: Indiana University Press.

Harré, R., and van Langenhove, L. (1999). The Dynamics of Social Episodes. In R. Harré and L. van Langenhove (Eds.), *Positioning Theory: Moral Contexts of Intentional Action*. (pp. 1-13) Oxford: Blackwell Publications.

Hart, C., Mulhall, P., Berry, A., Loughran, J., and Gunstone, R. (2000). What is the Purpose of this Experiment? Or Can Students Learn Something from Doing Experiments? *Journal of research in science teaching, 37*(7), 655-675.

Hasse, C. (1998). Learning to pattern physics virtues: male and female dissimilarities. In J. Höjgaard Jensen and M. Niss and T. Wedge (Eds.), *Justification and enrolment problems in education involving mathematics and physics*. (pp. 106-112) Roskilde: Roskilde University Press.

Hasse, C. (2002a). Gender Diversity in Play With Physics: The Problem of Premises for Participation in Activities. *Mind, culture, and activity, 9*(4), 250-269.

Hasse, C. (2002b). Learning Physical Space: The Social Designation of Institutional Culture. *Folk. Journal of the Danish Ethnographic Society, 44*, 171-194.

Hasse, C. (2006). Learning lab anthology. Learning through reactions - the social designation of institutional cultural code-curricula. In C. H. Sörensen (Ed.), *Body and learning: a transdiciplinary approach. An anthology*. Copenhagen: The Danish University of Education Press.

Hatchell, H. (1998). Girls' entry into higher secondary sciences. *Gender and Education, 10*(4), 375-386.

Havdala, R., and Ashkenazi, G. (2007). Coordination of Theory and Evidence: Effect of Epistemological Theories on Students' Laboratory Practice. *Journal of Research in Science Teaching, 44*(8), 1134-1159.

Hazari, Z., Tai, R. H., and Sadler, P. M. (2007). Gender differences in introductory university physics performance: The influence of high school physics preparation and affective factors. *Science Education, 91*, 847-876.

Heller, P., and Hollabaugh, M. (1992). Teaching problem solving through cooperative grouping. Part 2: Designing problems and structuring groups. *American Journal of Physics, 60*(7), 637-644.

Helm, H. (1980). Misconceptions in physics amongst South African students. *Physics Education, 15*(2), 92-97,105.

Henwood, F. (1996). WISE choices? Understanding occupational decision-making in a climate of equal opportunities for women in science and technology. *Gender and Education, 8*(2), 199-214.

Henwood, F. (1998). Engineering difference: discourses on gender, sexuality and work in a college of technology. *Gender and Education, 10*(1), 35-49.

Heron, P. R. L., and Meltzer, D. E. (2005). The future of physics education research: Intellectual challenges and practical concerns. *American Journal of Physics, 73*(5), 390-394.

Hestenes, D. (1996). *Modeling Methodology for Physics Teachers.* Paper presented at the International Conference on Undergraduate Physics Education, College Park.

Hestenes, D., and Wells, G. (1992). A mechanics baseline test. *The physics teacher, 30*, 159-162.

Hestenes, D., Wells, G., and Swackhamer, G. (1992). Force Concept Inventory. *The physics teacher, 30*, 141-153.

Hewson, P. W. (1982). A Case Study of Conceptual Change in Special Relativity: The Influence of Prior Knowledge in Learning. *European Journal of Science Education, 4*(1), 61-78.

Hey, V. (2006). The politics of performative resignification: translating Judith Butler's theoretical discourse and its potential for a sociology of education. *British Journal of Sociology of Education, 27*(4), 439-457.

Hildebrand, G. M. (2001). *Con/Testing Learning Models*, Paper presented at the National Association of Research in Science Teaching (NARST). St. Luis, MO.

Hoffman, L. (2002). Promoting girls' interest and achievement in physics classes for beginners. *Learning and Instruction, 12*(4), 447-465.

Hofstein, A., and Lunetta, V. N. (2003). The Laboratory in Science Education: Foundations for the Twenty-First Century. *Science Education, 88*(1), 28-54.

Honan, E., Knobel, M., Baker, C., and Davies, B. (2006). Producing Possible Hannahs: Theory and the Subject of Research. *Qualitative Inquiry, 6*(9), 9-32.

Häussler, P., Hoffman, L., Langeheine, R., Rost, J., and Sievers, K. (1998). A typology of students' interest in physics and the distribution of gender and age with each type. *International Journal of Science Education, 20*(2), 223-238.

Johnstone, A. H., Watt, A., and Zaman, T. U. (1998). The students' attitude and cognition change to a physics laboratory. *Physics Education, 33*(1), 22-29.

Johnstone, J., and Dunne, M. (1996). Revealing Assumptions: Problematising Research on Gender and Mathematics and Science Education. In L. H. Parker, L. J. Rennie and B. J. Fraser (Eds.), *Gender, Science and Mathematics. Shortening the Shadow*. (pp. 53-66) Dordrecht: Kluwer Academic Publishers.

Jones, A. T., and Kirk, C. M. (1990). Gender differences in students' interests in applications of school physics. *Physics Education, 25*, 308-313.

Jones, M. G., Howe, A., and Rua, M. J. (2000). Gender differences in students' experiences, interests, and attitudes toward science and scientists. *Science Education, 84*, 180-192.

Jorgensen, J. (2002). Engineering Selves. Negotiating Gender and Identity in Technical Work. *Management Communications Quarterly, 15*(3), 350-380.

Karelina, A., and Etkina, E. (2007). Acting like a physicist: Student approach study to experimental design. *Physical Review Special Topics - Physics Education Research, 3*(2), 1-12.

Klainin, S. (1988). Practical work and science education. In P. Fensham (Ed.), *Development and Dilemmas in Science Education*. (pp. 169-188) London: Falmer Press.

Kleinman, S. S. (1998). Overview of feminist pedagogy on the ideology of science. *Journal of research in science teaching, 35*(8), 837-844.

Koch, A. (2001). Training in metacognition and comprehension of physics texts. *Science Education, 85*, 758-768.

Kohl, P. B., and Finkelstein, N. D. (2008). Patterns of multiple representation use by experts and novices during physics problem solving. *Physical Review Special Topics - Physics Education Research, 4*, 1-13.

Kumashiro, K. K. (2001). 'Posts' Perspectives on Anti-Oppressive Education in Social Studies, English, Mathematics and Science Classrooms. *Educational Researcher, 30*(3), 3-12.

Kung, R. L. (2005). Teaching the concept of measurement: An example of a concept-based laboratory course. *American Journal of Physics, 73*(8), 771-777.

Kung, R. L., Danielsson, A., and Linder, C. (2005). *Metacognition in the student laboratory: Is increased metacognition necessarily better?* Paper presented at the EARLI, Nicosia, Cyprus.

Kung, R. L., and Linder, C. (2007). Metacognitive activity in the physics student laboratory: Is increased metacognition necessarily better? *Metacognition and Learning, 2*(1), 41-56.

Kvale, S. (1996). *InterViews: An Introduction to Qualitative Research Interviewing*. Thousand Oaks, CA: SAGE.

Kvande, E. (1999). 'In the belly of the beast' Constructing femininities in engineering organizations. *The European Journal of Women's Studies, 6*, 305-328.

Lakoff, G., and Johnson, M. (1980). *Metaphors We Live By*. Chicago: University of Chicago Press.

Larsen, K. M. (1995). Women in astronomy: Inclusion in introductory textbooks. *American Journal of Physics, 63*(2), 126-131.

Lave, J. (1988). *Cognition in Practice: Mind, Mathematics, and Culture in Everyday Life*. New York: Cambridge University Press.

Lave, J. (1992). *Learning as Participation in Communities of Practice.* Paper presented at the Annual Meeting of the American Educational Research Association, San Francisco, California.

Lave, J. (1993). The practice of learning. In S. Chaiklin and J. Lave (Eds.), *Understanding practice. Perspectives on activity and context.* (pp. 3-32) New York: Cambridge University Press.

Lave, J. (1996). Teaching, as Learning, in Practice. *Mind, Culture, and Activity, 3*(3), 149-164.

Lave, J., and Wenger, E. (1991). *Situated learning. Legitimate peripheral participation*. New York: Cambridge University Press.

Laws, P. W. (1997). Milikan Lecture 1996: Promoting active learning based on physics education research in introductory physics courses. *American Journal of Physics, 65*(1), 14-21.

Laws, P. W., Rosborough, P. J., and Poodry, F. (1999). Women's responses to an activity-based introductory physics program. *Physics Education Resources, American Journal of Physics, Suppl., 67*(7), S32-S37.

Lazarowitz, R., and Tamir, P. (1994). Research in using laboratory instruction in science. In D.L. Gabel (Ed) *Handbook of research on science teaching and learning*. (pp. 94-128) New York: Macmillan.

Letts, W. (2001). When Science is Strangely Alluring: interrogating the masculinist and heteronormative nature of primary school science. *Gender and Education, 13*(3), 261-274.

Lewin, K. (1952). *Field theory in social science*. London: Tavistock.

Light, K., Todd, D., DeHart, L., and Day, M. (2002). Women in Physics: A High School-College Interaction. *Physics Teacher, 40*, 531-532.

Lincoln, Y. S., and Guba, E. G. (1985). *Naturalistic inquiry*. Beverly Hills: Sage.

Linder, C., and Marshall, D. (1998). Linking physics students' development as independent and reflective learners with changes in their conceptions of science. In C. Rust (Ed.), *Improving student learning. Improving students as learners* (pp. 107-117). Oxford: Oxford Centre for Staff and Learning Development.

Linder, C., Östman, L., and Wickman, P.-O. (2007). *Promoting Scientific Literacy: Science Education in Transaction.* Paper presented at the Linnaeus Tercentenary Symposium, Uppsala.

Linder, C. J. (1992). Is teacher-reflected epistemology a source of conceptual difficulty in physics? *International Journal of Science Education, 14*(1), 111-121.

Linder, C. J. (1993). A Challenge to Conceptual Change. *Science Education, 77*(3), 293-300.

Linder, C. J., and Marshall, D. (1997). Introducing and evaluating metacognitive strategies in large class introductory physics teaching. In G. Gibbs and C. Rust (Eds.), *Improving Student Learning through Course Design* (pp. 411 - 420). Oxford: Oxonian Rewley Press.

Lindwall, O., and Lymer, G. (2008). The Dark Matter of Lab Work: Illuminating the Negotiation of Disciplined Perception in Mechanics. *Jounrla of the Learning Sciences, 17*(2), 180-224.

Linehan, C., and McCarthy, J. (2000). Positioning in Practice: Understanding Participation in the Social World. *Journal for the Theory of Social Behaviour, 30*(4), 435-453.

Lippman Kung, R. (2005). Teaching the concept of measurement: An example of a concept-based laboratory course. *American Journal of Physics, 73*(8), 771-777.

Lising, L., and Elby, A. (2005). The impact of epistemology on learning: A case study from introductory physics. *American Journal of Physics, 73*(4), 372-382.

Lorenzo, M., Crouch, C. H., and Mazur, E. (2006). Reducing the gender gap in physics education. *American Journal of Physics, 74*(2), 118-122.

Lundborg, A., and Schönning, K. (2007). *Maskrosfysiker - genusperspektiv på rekrytering, handledning och arbetsmiljö bland Uppsalas fysikdoktorander.* [Dandelion physicists – gender perspectives on recruitment, supervision and working environment among the PhD students in physics in Uppsala] Uppsala: Universitetstryckeriet.

Lykke, N. (2003). Intersektionalitet - ett användbart begrepp för genusforskningen. [Intersectionality – a useful concept in gender research] *Kvinnovetenskaplig tidskrift,* [Journal of women's studies] *24*(1), 47-57.

Mallow, J. V., and Hake, R. R. (2002). Gender Issues in Physics/science Education (GIPSE) -- Some Annotated References: National Science Foundation.

Maloney, D. P., O'Kuma, T. L., Hieggelke, C. J., and Van Heuvelen, A. (2001). Surveying students' conceptual knowledge of electricity and magnetism. *Physics Education Research, American Journal of Physics Suppl., 69*(7), S12-S23.

Mashhadi, A. (1994). Students' Conceptions of Quantum Physics. In C. Bernardini & C. Tarsitani and M. Vincentini (Eds.), *Thinking Physics for Teaching* (pp. 313-328). Rome: Plenum Press.

Maxwell, J. A. (1996). *Qualitative Research Design: An Interactive Approach.* Thousand Oaks, California: Sage.

McCloskey, M. (1983). Naive Theories of Motion. In D. Gentner and A. L. Stevens (Eds.), *Mental Models* (pp. 299-323). Hillsdale, NJ: Lawrence Erlbaum Associates.
McCullough, L. (2001). *A gender context for the Force Concept Inventory.* Paper presented at the National meeting for the American Association of Physics Teachers, San Diego, CA.
McCullough, L. (2002). Women in physics: A review. *The physics teacher, 40*, 86-91.
McCullough, L. (2004). Gender, context, and physics assessment. *Journal of International Women's Studies, 5*(4), 20-30.
McCullough, L. (2007). Gender in the Physics Classroom. *The Physics Teacher, 45*, 316-317.
McDermott, L. C. (1991). Millikan Lecture 1990: What we teach and what is learnt - Closing the gap. *American Journal of Physics, 59*(4), 301-315.
McDermott, L. C. (2001). Oerstedt Medal Lecture 2001: Physics Education Research - the key to student understanding. *American Journal of Physics, 69*(11), 1127-1137.
McDermott, L. C., and Redish, E. F. (1999). Resource Letter: PER-1: Physics Education Research. *American Journal of Physics, 67*(9), 755-767.
Mellström, U. (1995). *Engineering lives: Technology, time and space in a male-centred world.* Linköping: Linköping Studies in Art and Science.
Mellström, U. (1999). *Män och deras maskiner.* [Men and their machines] Nora: Bokförlaget Nya Doxa.
Mellström, U. (2002). Patriarchal machines and masculine embodiment. *Science, Technology, & Human Values, 27*(4), 460-478.
Mellström, U. (2004). Machines and masculine subjectivity. Technology as an integral part of men's life experience. *Men and Masculinities, 6*(4), 368-382.
Meltzer, D. (2005). Relation between students' problem-solving performance and representational format. *American Journal of Physics, 73*(5), 463-478.
Meltzer, D. E., and Manivannan, K. (2002). Transforming the lecture-hall environment: The fully interactive physics lecture. *American Journal of Physics, 70*(6), 639-654.
Menard, A., and Uzun, A. (1993). Educating women for success in physics: Lessons from Turkey. *American Journal of Physics, 61*(7), 611-615.
Merchant, C. (1984). Isis nya medvetenhet. [Isis' Consciousness Raised] *Kvinnovetenskaplig tidskrift,* [Journal of women's studies] *4.*
Millar, R., La Maréchal, J.-F., and Tiberghien, A. (1999). 'Mapping' the domain. Varieties of practical work. In J. Leach and A. Paulsen (Eds.), *Practical work in science education: Recent research studies* (pp. 33-59). Roskilde: Roskilde University Press.

Moore, E. (2006). 'You tell all the stories': Using narrative to explore hierarchy within a Community of Practice. *Journal of Sociolinguistics, 10*(5), 611-640.
Murphy, P. (1991). Gender differences in pupils' reactions to practical work. In B. Woolnough (Ed.), *Practical Science*. Buckingham: Open University Press.
Murphy, P., and Whitelegg, E. (2006a). *Girls in the Physics Classroom. A Review of the Research on the Participation of Girls in Physics*. London: Institute of Physics.
Murphy, P., and Whitelegg, E. (2006b). Girls and physics: continuing barriers to 'belonging'. *The Curriculum Journal, 17*(3), 281-305.
Müller, R., and Wiesner, H. (2002). Teaching quantum mechanics on an introductory level. *American Journal of Physics, 70*(3), 200-209.
Nair, I., and Majetich, S. (1995). Physics and Engineering in the classroom. In S. V. Rosser (Ed.), *Teaching the majority. Breaking the gender barrier in science, mathematics, and engineering*. (pp. 25-42) New York: Teachers College Press.
Nespor, J. (1994). *Knowledge in motion. Space, Time, and Curriculum in Undergraduate Physics and Management*. London: Falmer Press.
Norby, R. F. (2000). Equitable teaching of physics to women students: thoughts for the new decade. *The physics teacher, 38*, 494-495.
Northedge, A. (2002). Organizing excursions into specialist discourse communities: a sociocultural account of university teaching. In G. Wells and G. Claxton (Eds.), *Learning for life in the 21st century. Sociocultural perspectives on the future of education*. (pp. 252-264) Oxford: Blackwell Publishers.
Ormerod, M. B., Bottomley, J. M., and Keys, W. P. (1979). Girls and physics education. *Physics Education, 14*, 271-277.
Paechter, C. (2003a). Masculinities and femininities as communities of practice. *Women's studies international forum, 26*(1), 69-77.
Paechter, C. (2003b). Learning masculinities and femininities: power/knowledge and legitimate peripheral participation. *Women's studies international forum, 26*(6), 541-552.
Paechter, C. (2006b). Reconceptualizing the gendered body: learning and constructing masculinities and femininities in school. *Gender and Education, 18*(2), 121-135.
Paechter, C. (2006c). Power, knowledge and embodiment in communities of sex/gender practice. *Women's Studies International Forum, 29*(1), 13-26.
Paechter, C. (2007a). *What is it to construct and perform a masculine femininity? Untying masculinity from 'what men do'*. Paper presented at the 6th International Gender and Education Association Conference, Dublin, Ireland.
Paechter, C. (2007b). *Being boys. Being girls. Learning masculinities and femininities*. Poland: Open University Press.
Parker, K. (2002). Men are not from Mars. *Physics Education, 37*(1), 12-17.

Parsons, S. (1995). Making Sense of Students' Science: The Construction of a Model of Tinkering. *Research in Science Education, 25*(2), 203-219.
Perkins, K. K., Adams, W. K., Dubson, M., Finkelstein, N. D., Reid, S., Wieman, C. E., and LeMaster, R. (2006). PhET: Interactive Simulations for Teaching and Learning Physics. *The Physics Teacher, 44*, 18-23.
Petri, J., and Niedderer, H. (1998). A learning pathway in high-school level quantum atomic physics. *International Journal of Science Education, 20*(9), 1075-1088.
Phipps, A. (2002). Engineering women: the 'gendering' of professional identities. *International Journal of Engineering Education, 18*(4), 409-414.
Polkinghorne, D. E. (1995). Narrative configuration in qualitative research. In J. A. Hatch and R. Wisniewski (Eds.), *Life history and narrative* (pp. 5-23). London: Falmer Press.
Pollack, B. L., and Little, L. K. (1973). Experimental Project in Physics Education or New Avenues for Women. *The Physics Teacher*, 391-399.
Pollock, S. J., Finkelstein, N. D., and Kost, L. E. (2007). Reducing the gender gap in the physics classroom: How sufficient is interactive engagement? *Physical Review Special Topics - Physics Education Research, 3*, 1-4.
Posner, G., Strike, K., Hewson, P., and Gertzog, W. (1982). Accommodation of a scientific conception: Towards a theory of conceptual change. *Science Education, 66*(2), 211-227.
Redish, E. F. (1999). Milikan Lecture 1998: Building a Science of Teaching Physics. *American Journal of Physics, 67*(7), 562-573.
Redish, E. F. (2003). *Teaching physics with the physics suite*. Hoboken, NJ: John Wiley & Sons, Inc.
Redish, E. F. (2004). *A Theoretical Framework for Physics Education Research: Modeling Student Thinking*. Paper presented at the Enrico Fermi Summer School in Physics, Varenna, Italy.
Redish, E. F., Saul, J. M., and Steinberg, R. N. (1998). Student expectations in introductory physics. *American Journal of Physics, 66*(3), 212-224.
Reid, N. (2003). Gender and physics. *International Journal of Science Education, 25*(4), 509-536.
Richardson, L. (1994). Writing: a method of inquiry. In N. K. Denzin and Y. Lincoln (Eds.), *Handbook of qualitative research* (pp. 516-529). Thousand Oaks, CA: Sage.
Ritchie, S. M. (2002). Student Positioning within Groups During Science Activities. *Research in Science Education, 32*, 35-54.
Roberts, D. A. (2007a). Scientific Literacy/Science Literacy. In S. K. Abell and N. G. Lederman (Eds.), *Handbook of Research on Science Education*. (pp. 729-780) Mahwah, NJ: Lawrence Erlbaum Associates.

Roberts, D. A. (2007b). Linné Scientific Literacy Symposium Opening Remarks. In C. Linder, L. Östman and P.-O. Wickman (Eds.), *Promoting Scientific Literacy: Science Education in Transaction*. (pp. 9-17) Uppsala.

Robertson, L. A. (2006). Why are there so few female physicists? *The Physics Teacher, 44*, 177-180.

Robson, C. (2002). *Real world research* (Second edition). Padstow: Blackwell Publishing.

Rollnick, M., Allie, S., Buffler, A., Campbell, B., and Lubben, F. (2004). Development and application of a model for students' decision-making in laboratory work. *African Journal of Research in SMT Education, 8*(1), 13-27.

ROSE (Relevance of Science Education) (2009) Retrieved February 23, 2009 from http://www.ils.uio.no/english/rose/index.html

Rosser, S. V. (1995). Introduction: Reaching the Majority. Retaining Women in the Pipeline. In S. V. Rosser (Ed.), *Teaching the Majority. Breaking the gender barrier in science, mathematics, and engineering*. (pp. 1-25) New York: Teachers College Press.

Ryder, J., Leach, J., and Driver, R. (1999). Undergraduate science students' images of science. *Journal of research in science teaching, 36*(2), 201-219.

Salminen-Karlsson, M. (2006). Situating gender in situated learning. *Scandinavian Journal of Management, 22*(1), 31-48.

Scherr, R. E. (2007). Modeling student thinking: An example from special relativity. *American Journal of Physics, 75*(3), 272-280.

Schiebinger, L. (1991). *The mind has no sex? Women in the origins of modern science*. Cambridge MA: Harvard University Press.

Schneider, M. B. (2001). Encouragement of women physics majors at Grinnell college: A case study. *The physics teacher, 39*, 280-282.

Sencar, S., and Eryilmaz, A. (2004). Factors mediating the effect of gender on ninth-grade Turkish students' misconceptions concerning electric circuits. *Journal of Research in Science Teaching, 41*(6), 603-616.

Séré, M.-G. (2002). Towards Renewed Research Questions from the Outcomes of the European Project *Labwork in Science Education. Science Education, 86*, 624-644.

Séré, M.-G., Fernandez-Gonzalez, M., Gallego, J. A., Gonzalez-Garcia, F., De Manuel, E., Perales, F. J., and Leach, J. (2001). Images of Science Linked to Labwork: A Survey of Secondary School and University Students. *Research in Science Education, 31*(4), 499-523.

Seymour, E., and Hewitt, N. M. (1997). *Talking about leaving. Why Undergraduates Leave the Sciences*. Boulder: Westview press.

Sfard, A. (1998). On Two Metaphors for Learning and the Danger of Choosing Just One. *Educational Researcher, 27*(2), 4-13.

Shaffer, P. S., and McDermott, L. C. (1992). Research as a guide for curriculum development: An example from introductory electricity. Part II: Design of instructional strategies. *American Journal of Physics, 60*(11), 1003-1013.

Sherin, B. L. (2001). How students understand physics equations. *Cognitive Instruction, 19*(4), 479-541.
Sherr, R. E. (2008). Gesture analysis for physics education researchers. *Physical Review Special Topics - Physics Education Research, 4*, 1-9.
Singh, A., and Butler, P. H. (1990). Refraction: conceptions and knowledge structure. *International Journal of Science Education, 12*(4), 429-442.
Singh, C. (2008). Interactive learning tutorials on quantum mechanics. *American Journal of Physics, 76*(4&5), 400-405.
Sjøberg, S. (2000). *Naturvetenskap som allmänbildning. En kritisk ämnesdidaktik.* [Scientific literacy. Critical perspectives on science education] Lund: Studentlitteratur.
Skelton, C. (1998). Feminism and Research into Masculinities and Schooling. *Gender and Education, 10*(2), 217-227.
Smith III, J. P., diSessa, A., and Roschelle, J. (1993-1994). Misconceptions Reconceived: A Constructivist Analysis of Knowledge in Transition. *The Journal of Learning Sciences, 3*(2), 115-163.
Solomon, Y. (2007). Not belonging? What makes a functional learner identity in undergraduate mathematics? *Studies in Higher Education, 32*(1), 79-96.
Sowell, S. P. (2004). *Doing gender/teaching science: a feminist poststructural analysis of middle school science teachers' identity negotiations.* Unpublished PhD thesis, Florida State University.
Staberg, E.-M. (1994). Gender and science in the Swedish compulsory school. *Gender and Education, 6*(1), 35-45.
Stadler, H., Duit, R., and Benke, G. (2000). Do boys and girls understand physics differently? *Physics Education, 35*(6), 417-422.
Stake, R. E. (1994). Case Studies. In N. K. Denzin and Y. S. Lincoln (Eds.), *Handbook of qualitative research* (pp. 236-247). London: Sage.
Statistiska Centralbyrån (2008a) Statistiska meddelanden, UF 20 SM 0801. Serie Utbildning och Forskning. Retrieved March 2, 2009, from http://www.scb.se/statistik/UF/UF0205/2008A01/UF0205_2008A01_SM_UF20SM0801.pdf.
Statistiska Centralbyrån (2008b) Statistiska meddelanden, UF 23 SM 0801. Serie Utbildning och Forskning. Retrieved March 2, 2009, from http://www.scb.se/statistik/UF/UF0202/2008A02/UF0202_2008A02_SM_UF23SM0801.pdf.
Stewart, M. (1998). Gender issues in physics education. *Educational Research, 40*(3), 283-293.
Stonyer, H. (2002). Making engineering students - making women: the discursive context of engineering education. *International Journal of Engineering Education, 18*(4), 392-399.
Sweeney, A. E., and Paradis, J. A. (2004). Developing a laboratory model for the professional preparation of future science teachers: a situated cognition perspective. *Research in Science Education, 34*(2), 195-219.

Svennbeck, M. (2004). *Omsorg om naturen. Om NO-utbildningens selektiva traditioner med fokus på miljöfostran och genus.* [Care about nature. About the science education's selective traditions with a focus on environmental awareness and gender] Uppsala Studies in Education 104, Uppsala University, Uppsala.

Taber, K. (1991). Girl-friendly physics in the National Curriculum. *Physics Education, 26*, 221-226.

Tai, R. H., and Sadler, P. M. (2001). Gender differences in introductory undergraduate physics performance: university physics versus college physics in the USA. *International Journal of Science Education, 23*(10), 1017-1037.

Tan, E., and Calabrese Barton, A. (2007). Unpacking science for all through the lens of identities-in-practice: the stories of Amelia and Ginny. *Cultural Studies of Science Education, 3*(1), 43-71.

Taylor, J. (1979). Sexist bias in physics text books. *Physics Education, 14*, 277-280.

Thacker, B. A. (2003). Recent advances in classroom physics. *Reports on progress in physics, 66*, 1833-1864.

Thomas, K. (1990). *Gender and subject in higher education*. Buckingham: Open university press.

Thompson, N. (1979). Sex differentials in physics education. *Physics Education, 14*, 285-317.

Tolmie, A., and Howe, C. (1993). Gender and dialogue in secondary school physics. *Gender and Education, 5*(2), 191-208.

Traweek, S. (1988). *Beamtimes and Lifetimes. The world of high energy physicists*. Cambridge, MA: Harvard University Press.

Tsai, C.-H., Chen, H., Chou, C.-Y., and Lain, K.-D. (2007). Current as the key concept of Taiwanese students' understandings of electric circuits. *International Journal of Science Education, 29*(4), 483-496.

Udo, M. K., Ramsey, G. P., Reynolds-Alpert, S., and Mallow, J. V. (2001). Does physics teaching affect gender-based science anxiety? *Journal of Science Education and Technology, 10*(3), 237-247.

Wajcman, J. (1991). *Feminism confronts technology*. Cambridge: Polity Press.

Walford, G. (1981). Tracking down sexism in physics textbooks. *Physics Education, 16*, 261-265.

Walker, M. (2001). Engineering identities. *British Journal of Sociology of Education, 22*(1), 75-89.

Walkerdine, V. (1989). *Counting girls out*. London: Virago.

Walkerdine, V. (1990). *Schoolgirl fictions*. London: Verso.

van Aalst, J. (2000). An Introduction to Physics Education Research. *Canadian Journal of Physics, 78*, 57-71.

van Heuvelen, A. (1991). Learning to think like a physicist: A review of research-based instructional strategies. *American Journal of Physics, 59*(10), 891-897.

van Langenhove, L., and Harré, R. (1999). Introducing positioning theory. In R. Harré and L. van Langenhove (Eds.), *Positioning Theory: Moral*

Contexts of Intentional Action. (pp. 14-31) Oxford: Blackwell Publications.
Warren, J. W. (1979). *Understanding Force*. London: John Murray.
Weaver-Hightower, M. (2003). The "Boy Turn" in Research and Gender and Education. *Review of Educational Research, 73*(4), 471-498.
Wee, A. T. S., and Baaquie, B. E. (1993). Gender differences in undergraduate physics examination in performance and learning strategies in Singapore. *Physics Education, 28*, 158-163.
Wenger, E. (1998). *Communities of practice. Learning, meaning and identity.* Cambridge: Cambridge University Press.
Vetenskapsrådet [Swedish Research Council] (2002). Forskningsetiska principer inom humanistisk-samhällsvetenskaplig forskning. [Research ethics principles in humanistic-social scientific research] Vetenskapsrådet. Stockholm.
Whiteley, P. (1996). The gender balance of physics textbooks: Caribbean and British books, 1985-1991. *Physics Education, 31*, 169-174.
Wickman, P.-O. (2004). The Practical Epistemologies of the Classroom: A Study of Laboratory Work. *Science Education, 88*(3), 325-344.
Wickman, P.-O., and Östman, L. (2002). Learning as Discourse Change: A Sociocultural Mechanism. *Science Education, 86*(5), 601-623.
Wieman, C. E. (2002). Autobiography. In T. Frängsmyr (Ed.), *The Nobel Prizes 2001*. Stockholm: The Nobel Foundation.
Wieman, C. E., Perkins, K. K., and Adams, W. K. (2008). Oersted Medal Lecture 2007: Interactive simulations for teaching physics: What works, what doesn't, and why. *American Journal of Physics, 76*(4&5), 393-399.
Williams, C., Stanisstreet, M., Spall, K., Boyes, E., and Dickson, D. (2003). Why aren't secondary students interested in physics? *Physics Education, 38*(4), 324-329.
Williams, G. (2006). Sir, are you wearing nail varnish? *Physics Education, 41*(4), 317-322.
Willis, P. E. (1977). *Learning to Labour: How Working Class Kids Get Working Class Jobs*. Farnborough: Saxon House.
Wittmann, M., Steinberg, R. N., Redish, E. F. (1999). Making sense of how students make sense of mechanical waves. *The Physics Teacher, 37*(1), 15-21.
Volkwyn, T. S., Allie, S., Buffler, A., and Lubben, F. (2008). Impact of a conventional introductory laboratory course on the understanding of measurement. *Physical Review Special Topics - Physics Education Research, 4*, 1-10.
von Wright, M. (1999). *Genus och text. När kan man tala om jämställdhet I fysikläromedel?* [Gender and text. When can you talk about equality in physics textbooks?] Stockholm: Skolverket, Liber distribution.
Woolnough, J. A., and Cameron, R. S. (1991). Girls, boys and conceptual physics: An evaluation of a senior secondary physics course. *Research in Science Education, 21*(1), 337-344.

Yeo, S., and Zadnik, M. (2001). Introductory Thermal Concept Evaluation: Assessing Students' Understanding. *The physics teacher, 39*, 496-504.

Zohar, A. (2003). Her physics, his physics: gender issues on Israeli advanced physics classes. *International Journal of Science Education, 25*(2), 245-268.

Zohar, A., and Bronshtein, B. (2005). Physics teachers' knowledge and beliefs regarding girls' low participation in advanced physics classes. *International Journal of Science Education, 27*(1), 61-77.

APPENDICES

Appendix A: List of journals and search words in the PER and gender literature review

Journal/database	Search word(s)
Physics Education	gender, girl, women
Physics Teacher	gender, girl, women
American Journal of Physics	gender, girl, women
European Journal of Physics	gender, girl, women
Gender and Education	physics
Journal of Research in Science Teaching	gender AND physics
International Journal of Science Education	gender AND physics
Science Education	gender AND physics
Research in Science Education	gender AND physics
Physical Review Special Topic Physics Education Research	gender, girl, women
ERIC	gender AND physics

Appendix B: Excerpt from a laboratory instruction in Wave Optics

Vågoptik V 3

Uppgifter:

Planera och genomför en serie försök för att

1) bestämma bredden av en enkelspalt

2) bestämma avståndet mellan spalterna i en dubbelspalt

* 3) bestämma kvoten mellan spaltbredd och spaltavstånd för en dubbelspalt (hur gör man det enklast?)

D 4) jämföra teoretiska och experimentella intensitetskvoter mellan diffraktionsmaxima hos enkelspalten. Jämför gärna både teoretiska värden beräknade med hjälp av fasvektormodellen och med hjälp av Physics handbook.

D 5) studera trippelspalt och ev. kvadrupelspalt. Förklara diagrammets utseende med hjälp av fasvektormodellen; rita upp fasvektordiagram.

G 6) bestämma hålets diameter

G 7) bestämma gitterkonstanten hos reflexionsgittret

D = utförs bäst på datoruppställningen

G = utförs bäst på gamla uppställningen

* = extrauppgift

Vågoptik V 3

Uppgifter:

Planera och genomför en serie försök för att

1) bestämma bredden av en enkelspalt

2) bestämma avståndet mellan spalterna i en dubbelspalt

* 3) bestämma kvoten mellan spaltbredd och spaltavstånd för en dubbelspalt (hur gör man det enklast?)

D 4) jämföra teoretiska och experimentella intensitetskvoter mellan diffraktionsmaxima hos enkelspalten. Jämför gärna både teoretiska värden beräknade med hjälp av fasvektormodellen och med hjälp av Physics handbook.

D 5) studera trippelspalt och ev. kvadrupelspalt. Förklara diagrammets utseende med hjälp av fasvektormodellen; rita upp fasvektordiagram.

G 6) bestämma hålets diameter

G 7) bestämma gitterkonstanten hos reflexionsgittret

D = utförs bäst på datoruppställningen

G = utförs bäst på gamla uppställningen

* = extrauppgift

Appendix C: Interview protocol, the undergraduate students[1]

I. To start off the discussion/get the student to think about the education in a broader context:

- Why did you chose this education?
- What do you want to work with in the future? When did you decide that this was what you wanted to do? Why?

II. Laboratory work

The purpose of laboratory work: The teachers'

Your personal: What do you want/what do try to learn in the laboratory?

Learning of physics: How laboratory work contributes/doesn't contribute to this.

Laboratory work in a broader context:
Previous experiences/knowledge:

- Someone I interviewed earlier told me.. previous work… report writing – do you have any similar experiences?
- Are there any previous experiences/knowledge you wish you would have had?
- What are you skilled at? Can you connect that to something you've done earlier?
- What is it important to be skilled at, in order to gain as much as possible from physics labs?
- What would you recommend a friend who started to study physics to focus on in the lab?
- Are there any differences between what is need/what is good to bring to the lab compared to other parts of the physics education?

[1] Translated from Swedish.

If you had unlimited time… Differences between levels?

The future:

- What do you think you will remember from the laboratory work?
- Do you think you will use anything in your future working life?
- Could you imagine to work in a research laboratory, why/why not?
- Can you think of a lab exercise that was particularly useful? Why?
- Can you think of a lab exercise where you contributed with something special? In what way?

Several of the students I've interviewed have found it difficult to relate the laboratory work to a broader context. Why do you think that is?

'Some say that the student laboratory is a way for students to learn how physicists work, how the research community works.' – What do you think about this?

Norms and ideals:

- If you were to describe the ideal student, in particular in regard to working in the laboratory, how would they look? What experiences should they bring?
- What do you think about this ideal? Is it something you strive for, why/why not?
- If you were to convince the rest of the group to do something you've come up with, for example a method, how would you do this? What arguments are OK to use? Is it different at different levels? Would you do it differently in the laboratory versus for example when you solve problems in group?

Gender:

(Questions with a lot of room for interpretation – let the interviewee guide the discussion.)

- It's mainly guys who study physics, what role do you think that plays?
- Have you reflected on that you've chosen an education that many people view as traditionally male?
- If you think specifically about the lab, are there differences in how guys and girls work? Why do you think that is? Can it have to do with previous experiences?

- How does gender matter for you in physics? In the physics education? In the work in the laboratory?

Appendix D: Interview protocol, the graduate students[1]

I. 'Biographical' background data

a) Education. Why this one?
b) Age
c) Where they grow up
d) Work experience
e) Earlier studies – compare these to physics
f) Parents' occupations
g) Direction of physics studies
h) Future plans

II. Main interview

Theme: **Choice of subject**

a) Describe your Master's research project shortly. Why did you choose this ex-jobb?
   How did you chose the project?
b) Why did you choose to study physics?
c) Why did you choose to do experimental physics?

Theme: **The view of the physicist, how they see themselves as physicists**

d) Do you see yourself as a physicist? Or well on the way towards becoming a physicist? When did you start seeing yourself this way? Has this changed during the course of the Master's/PhD project? What is it that has contributed to this change? Are you a typical physicist? Why/why not?

Theme: **Learning to become a physicist**

e) Do you feel that your physics education has prepared you well for your current project? If not, what more would you have needed? How do you learn to become a physicist?

Theme: **View of knowledge**

f) What do you see as the purpose of laboratory work? In the lab you're currently working in versus the student laboratory?
g) How did you work in the student laboratory? (What parts of the lab work did you focus on? What did you see yourself as skilled at? How did you work together with your course-mates?) What have

[1] Translated from Swedish

you brought with you to your current project? What, according to you, are the main differences between working in the student laboratory and the laboratory you are working in now? Could you say something about the differences between physics and other subjects you've studied?

Theme: **Help and supervision**

h) When do you ask for help? Who do you ask for help? Who wouldn't you ask for help? How/why do you ask for help? Do you ask "differently" now compared to when you were taking course?
i) What have you been allowed to do in the lab? What haven't you been allowed to do? Has this changed during the course of the project? How has it changed?
j) Who is your supervisor? How is the supervision organised? Formal/informal supervision. Who is supervising you in practice? How do you experience having a female/male supervisor? Is that important to you? Would anything be different with a male/female supervisor? Why?

Theme: **Masculinity and femininity in physics**

k) How many men and women work in your research group? How do you think that affects the atmosphere in the group? The work you're doing? Why? Are there more Master's/PhD students in the group? Is this important? Why?
l) Is the proportion of men versus women in physics something you've reflected upon? How do you experience being a man/woman in physics?
m) What is masculinity and femininity to you? How is that expressed among physicists? Do you experience that the view of masculinity and femininity is different in the physics context compared to other contexts you have experience of? What is typical for male and female physicists?
n) Do you think role-models are important? Why? Do you need role-models? Do you have a role-model? Why?

Appendix E: Excerpts from the interviews with Kalle and Karin

Kalle

I: Du läser Naturvetarprogrammet, eller hur?

Kalle: Ja, matte/data.

[Stycke borttaget]

Kalle: Jag vet inte varför jag tog matte/data, dom sa att det spelar ingen roll, du får ändå läsa vad du vill sen...

I: Varför valde du Naturvetarprogrammet då, från första början?

Kalle: Jag hade sökt till KTH och Linköping och Chalmers då tror jag, sen låg jag typ tionde reserv tror jag, för mina gymnasiebetyg var inte direkt sådär... Jag [idrottade] på elitnivå gjorde jag på gymnasiet, tränade så mycket att man var... Och så läste jag gymnasiet tre mil ifrån min stad som jag kommer ifrån, så det var ju pendling då, så det var hem jämt och så i väg och träna, så man orkade aldrig plugga någonting. När skulle jag hinna plugga liksom. Så att...

I: Vad läst du på gymnasiet..?

Kalle: Först gick jag naturvetenskaplig då, men sen eftersom jag inte hann med och plugga någonting så var dom irriterade på mig då, så då läste jag industri sen... Verkstadsarbetare då, sen läst jag in på Komvux ett år allt som fattades...

I: Men varför blev det fysik då liksom..?

Kalle: Alltid tyckt att fysik är roligt. Men för min del har det alltid varit den experimentella delen, det har aldrig varit att man ska bli teoretiker eller något sådant...

I: Vad är det som lockar med det experimentella då?

Kalle: Öhm... Det är väl det här att... man får komma med mycket lösningar själv då, och så får man... tillverka dom här idéerna sen då, fast det är ju inte jag som får göra det då, utan det är ju dom i verkstan då... Men det är just det som lockar då, genom att det är så nära verkstadsarbete, är det egentligen...

I: Men du är inte så intresserad av att göra skruvandet själv, eller? Det är liksom att komma på lösningarna och sen dela iväg det till någon annan..?

Kalle: Jo, jag skulle gärna vilja göra allting själv, men jag får ju inte göra det om man säger så. Det är ju verkstan sen som tillverka då, det är ju lite synd. För det kan jag göra precis lika bra som dom.

I: Du skulle helst vilja vara med på hela kedjan..?

Kalle: Ja. Det är kul.

I: Vad vill du jobba med så småningom då?

Kalle: Öhm... Jag kommer ju inte fortsätta inom universitetsvärlden sen.

I: Varför inte?

Kalle: Det är ju stor chans att jag kommer att doktorera då. Men sen efter det tror jag inte att jag kommer att fortsätta, för att... jag är mer sugen att börja arbeta på något företag eller så, så att man får lite mer... ett stort företag då med resurser bakom sig, så att man kan då komma ut lätt i världen och så. Det kan man i och för sig göra på universitetet också, men först och främst har man ju lite mer betalt utanför och som det verkar nu är det ju rätt stor osäkerhet här inom universitetsvärlden, med pengar och allt sånt där...

I: Det är just den ekonomiska säkerheten som lockar med att jobba i industrin, eller är det arbetsuppgifterna också?

Kalle: Nej, det behöver det inte vara...

I: Det behöver väl kanske inte vara någon större skillnad...

Kalle: Då gör man nog mer vad man vill om man stannar kvar inom universitetsvärlden, om man är experimentalist då. Men... pengar lockar liksom. Det är väl det som avgör. Så är det lite större säkerhet också. Även om det är osäkert på arbetsmarknaden också, så är det större säkerhet än det är här.

[Stycke borttaget]

I: Men, just kurslab då, vad ser du som syftet med att ha labbar i fysik?

Kalle: Syfte med labbar i fysik är väl ... det jag får för mig i alla fall med syftet med en lab är, alla kanske inte förstår då genom att läraren eller föreläsaren eller vad vi ska kalla det står där framme och rabblar teori på teori, utan då är det meningen att man ska ha lite labbar då så att man får en förståelse för det hela, för hur det fungerar, typ som ett komplement till teorin...

I: Förståelse för hur det fungerar, kan du utveckla det lite grann..?

Kalle: Ja... men om man bara har föreläsningar, det passar ju inte alla liksom, vissa skiter i och gå och för vissa går det jättebra att bara sitta med och lyssna och sen har du ju dom här som både behöver föreläsningar och labbar eller bara labbar, så att man får jobba med själva ämnet då, så att dom får en förståelse av vad det handlar om och hur det fungerar då.

I: Vad funkar bäst för dig personligen då?

Kalle: Det är nog vilket som, det spelar inte så stor roll...

I: Men du känner inte att du behöver både och...?

Kalle: Det beror mycket på kursboken man har och föreläsaren givetvis, har man en riktigt bra kursbok då behöver man ju varken föreläsare eller lab egentligen för mig personligen, då kan jag läsa i boken och jag förstår liksom, är det såhär... Men som i en del ämnen då, som föreläsaren är inte så bra, och kursboken är bra typ efter du gått kursen och skrivit tentan och det, för då har du den kunskapen som behövs – då är ju labbarna riktigt bra då.

I: För att du får se det på ett sätt till, sen förhoppningsvis funkar något av de här sätten...

Kalle: Ja...

I: Vad innebär det för dig att lära dig fysik då, du säger att det funkar att läsa i boken eller sitta på föreläsningar eller så...

Kalle: Vad menar du mer att lära sig fysik..? Inlärning eller?

I: Ja, hur skulle du förklara vad det är att lära sig fysik? När har man lärt sig någonting..?

Kalle: När har man lärt sig någonting – det är egentligen när man kan beskriva det för någon som inte har förstått det, så att han förstår det, då har man lärt sig någonting. Så länge man inte kan det, då förstår man det egentligen inte, även om man kanske kan lösa problemet eller klarar tentan ändå, eller hur man nu vill uttrycka det...

I: Har du några tidigare erfarenheter så, som du känner har varit bra när du labbat, som har hjälpt dig när du labbat så..?

Kalle: Nja... Tidigare erfarenheter, jag vet inte...

I: Jag tänkte utanför fysiken här alltså... Att du läste verkstad på gymnasiet måste ju ha gjort att du inte har några större problem att hålla på och skruva på grejer...

Kalle: Det har man ju alltid... man vågar ju alltid, man är ju inte rädd för att det ska gå fel, sån har man ju blivit genom att man är verkstadsarbetare då, så man kopplar på då elektroniken, kan ju elen någorlunda då, för alla har ju gått en elkurs då. Det har man väl lite lättare för då.

I: Tror du att det är en skillnad gentemot dina kursare som har gått natur på gymnasiet och sen kommit hit direkt... Är dom räddare för att koppla in saker och så..?

Kalle: Det kan jag egentligen inte svara på om dom är då. Man har ju en bättre verklighetsuppfattning i alla fall tycker jag...

I: Hur då..?

Kalle: Om man har... Ofta då om man jämför med dom här som bara har gått natur så, inte säkert att dom har sommarjobbat och kommit ut i verkligenheten på samma sätt som verkstadsarbetare får göra då när man går på gymnasiet, man är ju alltid ute varje sommar, och företagen skriker för de vill ha någon som kan köra maskin. Ok, du är inte 18, men man får göra skitjobbet vid sidan om då. Så man har ju verkligen fått komma ut om man säger så, på olika företag och allt sånt där och se hur det är.

I: Det känner ju jag igen mig i också, jag har ju stått på pappersbruk varenda sommar...

[Stycke borttaget]

Kalle: Liksom, jämför man då med, så är det inte så vanligt att... visst folk sommarjobbar ju, men de kanske plockar jordgubbar då t.ex. jobbar på kommun då, ligger och solar hela sommarn, för det är ju vad dom gör...

I: Men att du har den här typen av erfarenheter, tror du att det formar det sätt du jobbar i labbet på?

Kalle: Ja, det tror jag – jag kan inte förklara hur, men jag skulle tippa på det. Man har... kanske inte dom vanliga kurslabbarna, men i alla fall nu när jag gjorde mitt examensarbete så kan jag nog tänka mig att jag jobbar på ett annat sätt jämfört med hur vanlig naturvetare jobbar om man säger så, speciellt eftersom man får skruva mycket själv och så. Man är inte... jag vet inte vad det är, men det känns som det är skillnad i alla fall, man behöver... man kan stå därnere helt själv och bara jobba liksom, man behöver inte ha någon hjälp direkt...

I: Du är mycket självständigare...

Kalle: Ja, jag skulle tippa på det, speciellt när det kommer till att jobba i själva labbet då...

I: Ja, det är väl mest på ex-jobbet du har haft den möjligheten.

Kalle: Ja.

I: Ser du någon skillnad på hur du har jobbat nu med ex-jobbet mot tidigare kurslab? Eller, vad är dom stora skillnaderna?

Kalle: Öh... Skillnaden är väl att det är mycket roligare att jobba i labbet om man gör sitt examensarbete jämfört med kurslab då, för kurslab har ju känt lite mer sådär beroende på vad man har för labass t.ex. då, att man bara känner att vad jobbigt det här är, vad dryg han är den där personen, vi blir aldrig klar liksom med den här labben.

[Stycket borttaget]

I: Men du känner dig generellt väldigt bekväm med att jobba i labbet..? Skulle kunna tänka dig att jobba själv också kanske i kurslab? Det skulle inte ha varit ett problem...

Kalle: Nej, det skulle det inte vart...

I: Men man jobbar alltid två och två..?

Kalle: Ja, inte i astronomin sen, då jobbar man ju själv.

I: Men i de tidiga kurserna, där man jobbar med någon, vad ska den – inte en person – men vad ska den vara bra på?

Kalle: Det ska ju vara någon som vill få labben gjord då. Om man säger så.

I: Att man blir klar och sen blir man godkänd och så får man gå, eller?

Kalle: Ja, det ska inte vara någon person som inte vill få den gjort om du förstår vad jag menar liksom…

I: Nej, inte riktigt…

Kalle: Men det här att - ja, nu förstår jag inte det här, nu förstår jag inte det här, nu får du förklara. Ja, hur ska man säga… Man vill helst jobba med någon som har gjort sina förberedande uppgifter och har någorlunda bra koll inom fysiken då, för annars jobbar du med någon som inte har gjort sina förberedande uppgifter och söker förståelse då och inte har förberett sig någonting då blir du aldrig klar, men blir bara frustrerad.

I: Någon som vill lite för mycket kanske? Någon som vill få ut för mycket av den här labben?

Kalle: Ja…

I: Det är ju rätt tidsbegränsat…

Kalle: Jag jobbade ju med en sån i kärn- och partikeln, det var ju fruktansvärt jobbigt.

I: Jaha, någon som var oförberedd eller..?

Kalle: Nej, men sökte för mycket förståelse då hela tiden. [ok] Då blir det bara frustrerande till slut.

I: Vad är det för typ av förståelse du söker då i labbet?

Kalle: Det beror ju lite på vad det är för lab då, vad det är för mening då. Men i kärn- och partikel då, då var det ju, då tittade man mest på radioaktiva material då skulle man ju, ja det här är farligt då liksom, man skulle få, det tog jag mer som en erfarenhet då, man skulle få liksom lite erfarenhet av när det kommer radioaktiva material, att det är farligt liksom. Man tittade då, ”ojdå, det här var det väldigt radioaktivt eller det här var inte så farligt då”.

I: Det viktigaste för dig med den labben var alltså att få prova på att jobba med de här lite farliga substanserna, det var inte så mycket att lära sig om sönderfall eller…

Kalle: Det gick man ju igenom innan då om man säger så, betasönderfall och alfa och allt det där, det var ju ganska lugnt så sett, utan det var mer att få känna på det lite.

[Stycke borttaget]

Karin

I: Men du doktorerar här alltså?

Karin: Mm, jag har väl precis börjat kan man väl nästan säga. Jag började i september, så jag är väl ganska ny, tycker jag fortfarande...

[Stycke borttaget]

Karin: Jag är civilingenjör, så att jag har läst teknisk fysik med materialvetenskap och sen, då gjorde jag mitt ex-jobb [på ett företag].

I: Var det det ex-jobbet handlade om också?

[Stycke borttaget]

I: Läste du grundutbildningen här i stan eller?

Karin: Mm, jag läste här på [universitetet] då.

I: Fram till ex-jobbet?

Karin: Ja, fram till ex-jobbet.

I: Ja. Hur kom det sig att du blev civilingenjör?

Karin: Det var lite av en slump

I: Mm. [båda skrattar]

Karin: För att, på gymnasiet så, min tanke då på civilingenjör var lite grann att det är bara de smarta som läser det.

I: Mm.

Karin: Och jag hade, jag hade väldigt bra betyg i gymnasiet för att jag var väldigt ambitiös.

I: Ja.

Karin: Öh, men jag tyckte inte att jag var tillräckligt smart, men så, när jag skulle liksom välja, jag var [utomlands] ett år efter gymnasiet och innan jag skulle välja till högskolan.

I: Ok.

Karin: Och, då funderade jag lite grann liksom, ja, men det var ju som, så därför sökte jag [ett ingenjörsprogram].

I: Mm.

Karin: För jag var intresserad av matte och jag ville definitivt läsa någonting naturinriktat, men mer liksom tillämpat kanske, mer tillämpat kanske, mer matematik än att bara läsa ett kemiprogram.

I: Mm.

Karin: Och då vart det civilingenjör som liksom passade bäst in på det jag ville göra, fast jag liksom inte hade tänkt att jag skulle bli civilingenjör.

I: Ok.

Karin: Förstod du hur jag menade?

I: Så det var inte titeln du var ute efter?

Karin: Det, var inte titeln, utan…

I: Mer innehållet i en utbildning…

Karin: Ja, typ. Så var det.

I: Så du gick natur på gymnasiet? Eller gick du teknisk?

Karin: Natur gick jag.

I: Varför valde du det?

Karin: Öh, då var det väl, det kändes ju som ett stort beslut då, men i efterhand kanske det inte spelar så stor roll. Men jag var väl intresserad av just natur och vetenskapsämnena liksom, redan på högstadiet och sen ville man ha en ganska bred bas att stå på, men liksom samhälle och historia var som inte lika intressant tyckte jag, det var mer flummigt för mig då. Här kan man säga att det är lite mer…

I: Vad jobbar dina föräldrar med förresten?

Karin: Öh, min pappa [arbetar med ekonomi], men han är pensionerad nu. Har då ett litet konsultföretag på sidan om. Min mamma är administratör på [företag].

I: Mm, har du någon anknytning till naturvetenskap eller teknik så där hemifrån?

Karin: Nej, inte. Nej. Det blev så ändå.

I: Hur kom det sig att du gick en så annorlunda väg då?

Karin: ... Mot vad mina föräldrar?

I: Ja.

Karin: Jag vet inte, jag tror det passade, eller det var som det jag var intresserad av och sen så, jag var inte... Pappa hade väl säkert varit intresserad av att jag läst ekonomi också, fast det var aldrig någon styrning hemifrån direkt, utan han var väl mer intresserad av att man skulle läsa på universitet, för det var väl bra. Och det var som aldrig någon fråga om, om eller inte man skulle läsa på universitet för mig, utan det var som lite bestämt.

I: Ok.

Karin: Men, så var det nog mer fallenhet, eller liksom vad jag var intresserad av som gjorde att jag valde...

I: Att du valde att börja doktorera så småningom då, hur gick det till?

Karin: Öhm... Så här kan man säga, jag har ju läst materialvetenskap då som civilingenjör och det är inte riktigt vad jag håller på med nu, det är ju relaterat, men det är inte... Materialvetarna blir ju mer mot stålindustrin eller nanomaterial eller beläggningstekniker eller sådana saker, medan jag håller på med [ett mer tvärvetenskapligt projekt] så att, när jag valde att börja doktorera så var det ganska självklart, det var mest att jag ville lära mig den delen som jag tycket att jag saknade, det som jag var intresserad av, som jag inte... Hade jag valt grundutbildning idag kanske jag hade valt något annat.

I: Ok.

Karin: Men jag är nöjd med det jag har läst. Men det var mer för att lära mig...

I: Vad skulle du ha valt idag då tror du?

Karin: Öhm, kanske energisystem, eventuellt. Eller…

I: Det fanns inte när du började kanske?

Karin: Jag tror att det var första året då som dom hade utbildningen.

I: Ja.

Karin: Och då var jag väl kanske lite skeptisk till en utbildning som inte har gått…

I: Ja, det är väl klart. Om, du kan säga något mer om ditt doktorand-projekt – vad handlar det om?

[Stycke borttaget]

I: Så det miljöperspektivet är viktigt för dig, eller?

Karin: Mm, energiperspektivet.

I: Energiperspektivet ja…

Karin: Tycker jag är intressant.

I: På vilket sätt?

Karin: Öhm… På vilket sätt?

[Stycke borttaget]

I: Så skulle du säga att du forskning är ganska tillämpad, eller?

Karin: Mm, jag tror att det, jag tycker att det är ganska tillämpat.

I: Och det var något om lockade, att det var…

Karin: Ja, för mig var det det. Så det är inte lika, öh, vad ska man säga, lika avancerat, eller inte lika tekniskt eller lika på detaljnivå kanske som andra som doktorerar på [min avdelning], t.ex. att då är det mer fysi-kaliskt, det här är mer ett systemperspektiv eller vad man ska säga.

I: Ok.

Karin: Förstår du hur jag..?

I: Ja, jag tror det... Så, om du kan säga något, vad är det du kommer att göra eller vad är det du gör liksom, när du forskar?

[Stycke borttaget]

I: Hur gick det till, alltså sökte du en doktorandtjänst eller? Hur hamnade du här som doktorand?

Karin: Mm, jo, jag sökte en tjänst som var utannonserad, men jag hade ju pratat med Gustav[1] då, min handledare. Han var även min ämnesgranskare på ex-jobbet och sedan efter det så började jag kolla lite vad, vad det fanns för doktorandtjänster då, och då var det ingenting som jag var direkt intresserad av eller inget som kändes. Och nu, så utannonserade dom flera doktorandtjänster och så sa han att den här kunde ju vara någonting, så att han skickade och sade att dom skulle annonsera ut och jag sökte på den tjänsten

I: Ja.

[Stycke borttaget]

I: Om du ser tillbaka på grundutbildningen, känner du att den förberett dig tillräckligt bra för liksom där du är nu?

Karin: Mmm... Jag tycker att, oftast att det känns som att man inte kan sånt som man borde kunna, det låter ju konst... men att man, att man känner igen saker, men man kan inte någonting i detalj känns det som. Man kan säga, "ja, men såhär, eller hur var det här, det här har jag ju läst, det här borde jag ju kunna". Men att det liksom, möjligtvis att den tiden som student var ju väldigt intensiv, att man läste mycket och att det klämdes ihop på väldigt kort tid, och det är ju lite av grejen också eller vad man ska säga. Men, och att man kan kolla upp och att man kan ta reda på, och att man har ett hum om hur man ska göra saker, men det är inte så att man kan saker utantill tycker jag

I: Nej... Är det någonting du skulle vilja ha haft... Eller skulle du vilja att grundutbildningen hade sett annorlunda ut?

Karin: Öh, det är svårt att säga, eventuellt att det hade varit bättre att läsa ett ämne i taget än att läsa flera kurser parallellt, men det är väl ock-

[1] Pseudonym

så hur det passar, för ibland så är det ju bra att läsa flera parallellt för man kan hoppa mellan, man blir inte lika, det blir inte lika ensidigt liksom, utan man kan... nu är jag less på att göra det här, då kan jag fokusera på det här ämnet liksom.

I: Mm.

Karin: Så det är nog lite både och, det är svårt att säga, ska man få in så pass mycket på kort tid, så...

I: Nej, det förstår jag... Om du skulle beskriva då för någon som inte är inne i den här världen som vi är inne i, hur du lär dig ditt yrke, eller hur du har lärt dig ditt yrke, hur skulle du beskriva det?

Karin: Öh [skratt] Jag vet inte om jag kan ett yrke, det känns, jag känns som jag fortfarande lär mig att lära mig, eller det kanske, jag tycke inte, det är ju ett jobb, alltså, jag kommer ju hit och det är ett jobb liksom, men det är inte att nu måste jag göra A, B och C idag i den här ordningen, utan... Och det tycker jag är väldigt bra. Hur lär man sig att vara doktorand, eller det är det som är frågan?

I: Mm.

Karin: Jag tror att det är mycket... Att man måste vara, ta mycket eget initiativ liksom, att man vill, vad man vill göra och hur man vill göra det och lägga fram egna idéer och man måste ju naturligtvis ha viss bakgrund, liksom, rent kunskapsmässigt, men att, det behöver inte vara relaterat till vad man har för betyg egentligen tycker jag, utan mer vad man har lärt dig och inställningen till lärandet, tror jag, att man vill lära sig mer och att det man inte kan får man fylla på med liksom.

I: Hur ser du på din yrkesroll idag? Hur skulle du definiera dig själv liksom?

Karin: Eh [skratt] du tänker, skulle jag kalla mig själv student eller?

I: Skulle du kalla dig civilingenjör, skulle du kalla dig fysiker, skulle du kalla dig dokt...

Karin: Doktorand skulle jag kalla mig.

I: Ser du på dig själv som fysiker?

Karin: Nej, det gör jag inte, tror jag.

I: Varför inte?

Karin: Öh, för jag tycker inte att jag har läst, jag har väl läst fysik, men inte som en fysiker, känns mer smalt. Jag har läst så mycket annat också, så jag tycker inte att jag är tillräckligt spets i fysik liksom.

I: Vem skulle du definiera som fysiker?

Karin: ... Öh, så här spontant... så... en professor på våran avdelning som precis ska gå i pension liksom, han känns som att han är fysiker, han har gått den gamla skolan med fysik och sen doktorerat och så vidare. Det är som, den spontana tanken på en fysiker, och jag vet inte om det är... Att det känns mer som äldre män. Vilket låter ju jättetrist att man känner så, men...

I: Mm.

Karin För att, andra som kommer, typ en gymnasielärare jag hade i fysik då, han var också fysiker för mig för han var så, ja...

I: Så hur är den typiske fysikern? Vad har den för egenskaper och så vidare?

Karin: Sen måste jag ju också säga, att jag tycker att [en doktorandkollega] t.ex. är ju fysiker, hon har läst fysikprogrammet och hon är ju mer intresserad av, liksom, saker på en helt annan nivå än vad jag är, jag ser liksom till stora systemperspektivet och hon är mer intresserad av bandgap och hur elektroner hoppar, mer på den nivån. Vad var andra frågan? Hur beskriver jag en fysiker?

I: En typisk fysiker, hur är en typisk fysiker, vad har den för egenskaper t.ex.?

Karin: Öhm, måste fundera lite. Jag tror, intresserad av att ta reda på hur saker och ting fungerar. Både forskningsmässigt, men också i vardagen så finns det ju fysik liksom överallt, att vilja förklara saker ur ett fysiskt perspektiv.

I: Är det någonting du känner igen dig i?

Karin: Mm, det kan jag väl känna igen, att man vill veta hur saker fungerar.

[Stycke borttaget]

I: Ja, experimentell fysik pratade vi ju om, jag flaggade ju för att jag är intresserad av labbar liksom, så även om du inte har läst en fysikutbildning så, du har ju läst en hel del fysik och jag antar att du har labbat en del i fysikkursen och sådär

Karin: Mm.

I: Men, mer generellt, vad skulle du säga att syftet är med att ha labbar i en utbildning.

Karin: Öh, jag tycker oftast att man för större förståelse för någonting som man inte skulle förstå om man läste det i en bok, eller att man, att det blir förklarat på ett annat sätt, att man får se det liksom och göra istället för att bara läsa, eller bara höra.

I: Det var det syfte som labbar under din utbildning fyllde för dig?

Karin: Oftast när man tänker tillbaka så känns det som man kan komma ihåg saker som man har gjort laborationer på bättre än sånt som man bara har läst tycker jag.

I: Mm, det kan jag tänka mig… Vad är det för skillnad om du ser till kurslab under utbildningen kontra lab du jobbar i nu eller som du jobbade i som ex-jobbare?

Karin: Det är mindre styrt, för kurslab är ju oftast, gör A, gör B, gör C lite grann, att det är väldigt, för man har ju bara så mycket tid och man måste ändå komma fram till någonting på den där tiden. Så det är mycket mer styrt, medan när man labbar, både som jag gjorde i [utomlands] och som jag har gjort här så är det mer liksom, vi provar det här, äh, prova det här istället. Så tycker jag kanske att, det är mer att det finns inget facit eller det finns ingen gång för hur man ska göra utan man får liksom försöka hitta den själv.

I: Vad har du tagit med dig, har du tagit med dig någonting från kurslabbet till liksom att jobba i lab nu?

[Stycke borttaget]

I: När du tänker tillbaka på kurslab då, hur, hur jobbade du där, vad såg du dig själv som bra på t.ex. när du kurslabbade och hur jobbade du tillsammans med dina kursare och så där?

Karin: Mm… noggrann och strukturerad, det var som mina… för att jag tyckte att det var bättre att inte hasta i väg, utan att man gör den liksom, man tar sina mätpunkter eller vad det nu är man ska göra, att man liksom – noggrann och strukturerad är väl som sammanfattningen.

I: Vad fokuserade du på, vilka delar av arbetet i kurslab? Man kan ju tänka sig, det är ju förberedelser och det är ju det experimentella och det finns ju efterarbete och... säkert andra saker också.

Karin: Mm... Svårt att säga vad man fokuserade mest på. Jag tyckte att det var viktigt att få mätdata som man endera kan förklara varför det inte är rätt, eller varför vart det en avvikelse här som vi inte har tänkt på eller att man kan förklara hur det fungerar och varför det är rätt eller vad man förklara det med teorin liksom.

I: Mm.

Karin: Tror jag.

I: Vad skulle du säga att de viktigaste skillnaderna är mellan kurslab och ett forskningslab?

Karin: ... Oj, det var svårt. ... Att det är mer styrt igen, tror jag, jag kanske missförstår...

I: Nej, visst inte.

Karin: Att det är mer styrt och så att det är ett mindre område, att det är lite mer utsvävande och så kanske.

I: Det var typ samma fråga [båda skrattar] jag var bara intresserad av en utveckling eventuellt.

Karin: Eventuellt en utveckling, det var ju lite finurligt då, att ställa samma fråga igen [båda skrattar]. Ja...

I: Det var inte meningen att vara finurlig.

Karin: Jag vet inte om jag kan säga så mycket mer.

I: Nej, vi kan gå vidare. Jag tänkte att vi skulle prata lite om handledning och så också.

Karin: Ja.

I: För det tycker jag är intressant. [...] Vem har du som handledare?

Karin: Gustav Andersson har jag som handledare.

I: Hur, hur funkar handledningen, hur är det organiserat?

Karin: Öh, det är... det här är ju som ganska intressant då [Stycke borttaget] Öh, och sen så tror jag att, man... Jag tycker att jag kan lägga, jag kan fråga vad jag vill och jag kan lägga fram vad jag tycker är ett problem, men sen så... Det finns ju alltid brister eller saker man skulle vilja vara annorlunda, så tror jag det är för de flesta doktorander, det är inte alla som är tipptopp nöjd med sin handledare liksom, utan det finns alltid något man kan klaga på. Men, nu har jag inte varit så länge. Men vad jag kan se som skulle kunna vara ett problem är att inte, han har ganska lite tid, alltså, och det tror jag är ganska genomgående för många som är handledare, att dom har mycket annat liksom. Och, att, särskilt nu i början skulle jag vilja ha lite mer handfast styrning liksom, att, ja, men, jag tycker inte riktigt att jag har ett projekt än. Alltså, jag tänker mig ett projekt som någonting som kan leda till en publikation eller en artikel liksom och jag tycker inte att jag har börjat med det än och det är väl kanske inte så konstig. Jag har ju lite som jag måste läsa in mig på, som jag måste träna på och såna saker, så det... Men det kan ändå bli lite stressande och det kan väl liksom att... när jag är frustrerad och har problem så pratar jag med Gustav och så säger jag att så här är det, och det här tycker jag är jobbigt och så säger han någonting lite så halvflummigt tillbaka och så tycker jag att det känns bättre, men egentligen så har vi inte löst problemet [skratt] Det är lite.

I: Vad brukar problemet vara då?

Karin: Ibland så är ju problemet bara att man har kört fast och att man, det räcker att man får säga det själv, vad som är problemet, så när man har definierat problemet och berättat det för någon annan så kommer man på lösningen själv. Men i stora drag så är det ju att, jag skulle vilja ha mer handfast, i alla fall nu i början, och som doktorand ska man ju hitta på saker själv liksom, och det förstår jag också, men jag vill ha någonting att komma igång med liksom.

I: Är det du som definierar ditt eget projekt? Eller? Hur funkar det?

Karin: Äh, det är ganska lulligt, eller luddigt förklarat då mitt projekt, eftersom det har liksom, väldigt system, det är väldigt stort, det är inte, nu ska vi kolla på den här atomen och hur den rör sig hit liksom, det är som lite mer flummigt, eller vad ska man säga. Och när jag tog tjänsten var det väl, du ska kolla på [område] och det är ju som ett ganska stort område så att… och sen får jag göra, det är ju positivt också, jag får ju göra lite vad jag vill och styra mycket så, så det tycker jag är positivt, men…

I: Så det är samtidigt någonting du trivs med?

Karin: Ja.

I: Att få göra det.

Karin: Mm, jag är… mm.. Jag tycker om att göra lite som jag vill kan man väl säga.

I: Har ni liksom såhär formellt organiserad handledning eller är det mer såhär informellt, man träffas i korridoren och liksom..?

Karin: Både och. Det är ganska informellt eftersom vi, vi träffas ju nästan varje dag, och då kan man ju säga, eller jag brukar säga, "har du tid att prata på onsdag för nu vill jag, nu ska jag göra klart det här" och då vill jag liksom diskutera det här och då bestämmer vi en tid och säger att, nu vill jag prata om det här liksom. Så det är väl lite både och, för samtidigt kan jag bara komma förbi och säga, "nu har jag gjort det här, har du tid?".

I: Ok. Är Gustav den som du ber om hjälp eller finns det andra personer där också som kan… hjälpa till?

Karin: Det beror på, vad det är jag vill ha hjälp med, jag kan nog fråga andra också, beroende på vad det är. Nu är det ju så att det är ingen som håller på med det jag håller på med, så att jag känner, det är ju svårt för mig att fråga om hjälp. [Stycke borttaget]

I: Mm.

Karin: Alltså, det är många som kan mycket, men inte precis.

I: Så du är ganska ensam i ditt specifika projekt?

Karin: Ja, jag kan känna så ibland.

I: Finns det fler doktorander knutna till det projektet?

[Stycke borttaget]

I: Det är mycket enklare antar jag, att inte behöva förklara hela bakgrunden varje gång... När ber du om hjälp förresten?

Karin: Öhm... Jag vet själv att jag kan vara ganska dålig på att be om hjälp.

I: Ok.

Karin: För att jag, för det första så, är jag envis och jag vill lösa problemet själv [skratt] och det är ju ganska fjantigt, men jag vill liksom veta att jag har gjort allt jag kan på det här innan, och det är inte för att jag liksom är rädd att säga att det här kan jag inte, utan det är mer att jag är envis liksom, att, det här borde ju gå, lite grann så.

I: Hur tror du det kommer sig då?

Karin: [skratt] Ja, jag vet inte, jag har nog alltid varit envis tror jag. Kan själv grejen var ju som det första jag lärde mig säga tror jag, ah... [skratt] och sen så... ibland så kan jag tycka att jag inte vill störa andra. Lite så. Men är det någonting jag verkligen kör fast med så... är det nog lättare för mig att säga, kan vi ta det här någon gång när det passar dig, vilken... att man, så att man inte känner att man tränger sig på liksom.

I: Vem ber du om hjälp i första hand? Är det Gustav, eller?

Karin: Mm, det är nog mycket Gustav faktiskt.

I: Finns det andra personer som du inte skulle gå till?

Karin: Ja...

I: Och varför då i så fall?

Karin: Öh... det är nog ingen som jag känner som... eller, vad ska man svara på det. Om det är någon som, alltså, vi håller ju på med väldigt olika grejer och är det någon som jag vet inte har en susning om mitt problem är det som värdelöst att fråga den personer.

I: Jo, jo.

Karin: Men om man säger, att man skulle anta att alla personer hade den kunskap som behövdes så är det inget personligt som jag känner att jag inte skulle kunna fråga.

I: Nej.

Karin: Förstod du?

I: Ja, jag förstår. Nej, men man kan ju ha en massa olika anledningar till att man inte, ja, frågar en viss person så där...

Karin: Nej, jag tror inte... Jag tycker att det är förvånansvärt bra faktiskt, eller vad man ska säga...

I: Upplever du att du frågar annorlunda nu jämfört när du var på grundutbildningen, på andra sätt eller av andra orsaker?

Karin: Mm... Öh, jag eventuellt... Jag tror att på grundutbildningen så kanske man frågade mer för att förstå... förstå det specifika problemet kanske, medans nu så vill man mer lära sig att större, ett större område, eller bakgrunden på ett annat sätt, medan då var man mer fokuserad på hur ska jag förstå det här problemet så att jag klarar tentan.

[Stycke borttaget]

I: I den här forskningsmiljön du rör dig i, ungefär hur är andelen män och kvinnor där?

Karin: Öh... [skrattar till] Som doktorand, om man räknar doktoranderna så är det fifty-fifty män och kvinnor.

I: Ja.

Karin: Som professorer är det ju övervägande män. Jag tror vi har en kvinnlig professor

I: Ok.

Karin: That's it. Men på doktorandnivå tycker jag det är fifty-fifty...

I: Hur tror du det påverkar atmosfären? Eller sättet ni jobbar på?

Karin: Mm... öh... Jag tycker inte att det är en speciellt grabbigt atmosfär normalt sett

I: Nej.

Karin: Eftersom det är så många tjejer som doktorerar att det blir liksom en balans trots att många av professorerna är manliga. Öhm... däremot så... så ibland kan jag tycka att om det är någonting som ska göras, om man ska dra ihop till en fest eller om man ska fixa så att, någon fyller år så man ska samla in pengar, så tycker jag att det är oftare att den kvinnliga delen tillfrågas att ordna det här, men det är ju som inte arbetsrelaterat direkt, utan... än den manliga.

I: Är liksom andelen män och kvinnor i fysikmiljön, brett definierat, är det någonting du har reflekterat över under din utbildning eller nu eller sådär?

Karin: Mm, det tror jag, jag reflekterade ju över att det var ganska mycket tjejer när jag började här, eller liksom att man har tänkt, det har ju vi pratat om liksom, man har tänkt tanken, det är inte så att ojdå. Du förstår. Man har i alla fall noterat det. Sen, på grundutbildningen var det ju också ganska, där var det kanske lite grabbigare atmosfär, för du var vi ju... sex tjejer som mest liksom.

I: Av hur många?

Karin: 22, 24 kanske. Då är det ju mestadels killar. Och där vet jag, det tyckte jag var lite intressant själv faktiskt, tjejer, då tyckte vi att vi kunde ha tjejkväll liksom, för vi var ju ändå bara sex stycken. Och det är ganska, det blir ganska grabbigt och det är trevligt förvisso, liksom, men ibland kan det vara skönt att bara vara tjejerna, för då blir vi hörda liksom, och det var faktiskt killar som blev jätteupprörda [skratt] över att vi hade tjejkväll.

I: Jaha.

Karin: På grundutbildningen, för man tyckte inte att man fick utesluta en grupp på så sätt.

I: Nähä. Hur tog den här grabbigheten sig uttryck då?

Karin: Öh, kanske lite vad man diskuterade, att man diskuterade mer, teknik. Det var ju inte så att, områden man pratade om kanske , vi pratade om helt andra saker när vi var bara tjejerna och att killarna hördes nog mer oftast än tjejerna. Det var inte så att vi inte, det är nog bara som det faller sig lite grann, för det var inte så att vi inte pratade, eller att tjejerna satt helt tysta i ett hörn liksom, för vi hade ju bra sammanhållning allihopa, men... Alltså, jag trivs, jag trivdes jättebra, men det vart annorlunda när bara tjejerna träffades och det är väl kanske ganska normalt, men...

I: Vad var det som var annorlunda då, förutom att ni pratade om andra saker?

Karin: Mer stillsamt tror jag. Fast det är kanske vad man gör också, kanske, jag vet inte... Och sen, ja, det är svårt...

I: Hur har du upplevt det att vara kvinna i fysikmiljön?

Karin: Jag tycker att det har varit bra faktiskt. Alltså jag tycker att man har fått positiv respons för att man är tjej, det tycker jag, att man kanske blir särbehandlad på ett positivt sätt, på något sätt, eller att man blir peppad på positivt sätt, eller, jag vet inte... Ja. Jag tycker att det har varit ganska bra.

[Stycke borttaget]

Acta Universitatis Upsaliensis

Uppsala Dissertations from the Faculty of Science

Editor: The Dean of the Faculty of Science

1–11: 1970–1975

12 *Lars Thofelt:* Studies on leaf temperature recorded by direct measurement and by thermography. 1975.

13 *Monica Henricsson:* Nutritional studies on Chara globularis Thuill., Chara zeylanica Willd., and Chara haitensis Turpin. 1976.

14 *Göran Kloow:* Studies on Regenerated Cellulose by the Fluorescence Depolarization Technique. 1976.

15 *Carl-Magnus Backman:* A High Pressure Study of the Photolytic Decomposition of Azoethane and Propionyl Peroxide. 1976.

16 *Lennart Källströmer:* The significance of biotin and certain monosaccharides for the growth of Aspergillus niger on rhamnose medium at elevated temperature. 1977.

17 *Staffan Renlund:* Identification of Oxytocin and Vasopressin in the Bovine Adenohypophysis. 1978.

18 *Bengt Finnström:* Effects of pH, Ionic Strength and Light Intensity on the Flash Photolysis of L-tryptophan. 1978.

19 *Thomas C. Amu:* Diffusion in Dilute Solutions: An Experimental Study with Special Reference to the Effect of Size and Shape of Solute and Solvent Molecules. 1978.

20 *Lars Tegnér:* A Flash Photolysis Study of the Thermal Cis-Trans Isomerization of Some Aromatic Schiff Bases in Solution. 1979.

21 *Stig Tormod:* A High-Speed Stopped Flow Laser Light Scattering Apparatus and its Application in a Study of Conformational Changes in Bovine Serum Albumin. 1985.

22 *Björn Varnestig:* Coulomb Excitation of Rotational Nuclei. 1987.

23 *Frans Lettenström:* A study of nuclear effects in deep inelastic muon scattering. 1988.

24 *Göran Ericsson:* Production of Heavy Hypernuclei in Antiproton Annihilation. Study of their decay in the fission channel. 1988.

25 *Fang Peng:* The Geopotential: Modelling Techniques and Physical Implications with Case Studies in the South and East China Sea and Fennoscandia. 1989.

26 *Md. Anowar Hossain:* Seismic Refraction Studies in the Baltic Shield along the Fennolora Profile. 1989.

27 *Lars Erik Svensson:* Coulomb Excitation of Vibrational Nuclei. 1989.

28 *Bengt Carlsson:* Digital differentiating filters and model based fault detection. 1989.

29 *Alexander Edgar Kavka:* Coulomb Excitation. Analytical Methods and Experimental Results on even Selenium Nuclei. 1989.

30 *Christopher Juhlin:* Seismic Attenuation, Shear Wave Anisotropy and Some Aspects of Fracturing in the Crystalline Rock of the Siljan Ring Area, Central Sweden. 1990.

31 *Torbjörn Wigren:* Recursive Identification Based on the Nonlinear Wiener Model. 1990.

32 *Kjell Janson:* Experimental investigations of the proton and deuteron structure functions. 1991.

33 *Suzanne W. Harris:* Positive Muons in Crystalline and Amorphous Solids. 1991.

34 *Jan Blomgren:* Experimental Studies of Giant Resonances in Medium-Weight Spherical Nuclei. 1991.

35 *Jonas Lindgren:* Waveform Inversion of Seismic Reflection Data through Local Optimisation Methods. 1992.

36 *Liqi Fang:* Dynamic Light Scattering from Polymer Gels and Semidilute Solutions. 1992.

37 *Raymond Munier:* Segmentation, Fragmentation and Jostling of the Baltic Shield with Time. 1993.

Prior to January 1994, the series was called *Uppsala Dissertations from the Faculty of Science.*

Acta Universitatis Upsaliensis

Uppsala Dissertations from the Faculty of Science and Technology

Editor: The Dean of the Faculty of Science

1–14: 1994–1997. 15–21: 1998–1999. 22–35: 2000–2001. 36–51: 2002–2003.

52. *Erik Larsson:* Identification of Stochastic Continuous-time Systems. Algorithms, Irregular Sampling and Cramér-Rao Bounds. 2004.
53. *Per Åhgren:* On System Identification and Acoustic Echo Cancellation. 2004.
54. *Felix Wehrmann:* On Modelling Nonlinear Variation in Discrete Appearances of Objects. 2004.
55. *Peter S. Hammerstein:* Stochastic Resonance and Noise-Assisted Signal Transfer. On Coupling-Effects of Stochastic Resonators and Spectral Optimization of Fluctuations in Random Network Switches. 2004.
56. *Esteban Damián Avendaño Soto:* Electrochromism in Nickel-based Oxides. Coloration Mechanisms and Optimization of Sputter-deposited Thin Films. 2004.
57. *Jenny Öhman Persson:* The Obvious & The Essential. Interpreting Software Development & Organizational Change. 2004.
58. *Chariklia Rouki:* Experimental Studies of the Synthesis and the Survival Probability of Transactinides. 2004.

59. *Emad Abd-Elrady:* Nonlinear Approaches to Periodic Signal Modeling. 2005.
60. *Marcus Nilsson:* Regular Model Checking. 2005.
61. *Pritha Mahata:* Model Checking Parameterized Timed Systems. 2005.
62. *Anders Berglund:* Learning computer systems in a distributed project course: The what, why, how and where. 2005.
63. *Barbara Piechocinska:* Physics from Wholeness. Dynamical Totality as a Conceptual Foundation for Physical Theories. 2005.
64. *Pär Samuelsson:* Control of Nitrogen Removal in Activated Sludge Processes. 2005.
65. *Mats Ekman:* Modeling and Control of Bilinear Systems. Application to the Activated Sludge Process. 2005.
66. *Milena Ivanova:* Scalable Scientific Stream Query Processing. 2005.
67. *Zoran Radovic':* Software Techniques for Distributed Shared Memory. 2005.
68. *Richard Abrahamsson:* Estimation Problems in Array Signal Processing, System Identification, and Radar Imagery. 2006.
69. *Fredrik Robelius:* Giant Oil Fields – The Highway to Oil. Giant Oil Fields and their Importance for Future Oil Production. 2007.
70. *Anna Davour*: Search for low mass WIMPs with the AMANDA neutrino telescope. 2007.
71. *Magnus Ågren*: Set Constraints for Local Search. 2007.
72. *Ahmed Rezine*: Parameterized Systems: Generalizing and Simplifying Automatic Verification. 2008.
73. *Linda Brus*: Nonlinear Identification and Control with Solar Energy Applications. 2008.
74. *Peter Nauclér:* Estimation and Control of Resonant Systems with Stochastic Disturbances. 2008.
75. *Johan Petrini:* Querying RDF Schema Views of Relational Databases. 2008.
76. *Noomene Ben Henda:* Infinite-state Stochastic and Parameterized Systems. 2008.
77. *Samson Keleta:* Double Pion Production in $dd \rightarrow \alpha\pi\pi$ Reaction. 2008.
78. *Mei Hong:* Analysis of Some Methods for Identifying Dynamic Errors-in-variables Systems. 2008.
79. *Robin Strand:* Distance Functions and Image Processing on Point-Lattices With Focus on the 3D Face-and Body-centered Cubic Grids. 2008.
80. *Ruslan Fomkin:* Optimization and Execution of Complex Scientific Queries. 2009.
81. *John Airey:* Science, Language and Literacy. Case Studies of Learning in Swedish University Physics. 2009.
82. *Arvid Pohl:* Search for Subrelativistic Particles with the AMANDA Neutrino Telescope. 2009.

83. *Anna Danielsson:* Doing Physics – Doing Gender. An Exploration of Physics Students' Identity Constitution in the Context of Laboratory Work. 2009.